CW00493764

WORKPLACE HEALTH & SAFETY IN EUROPE

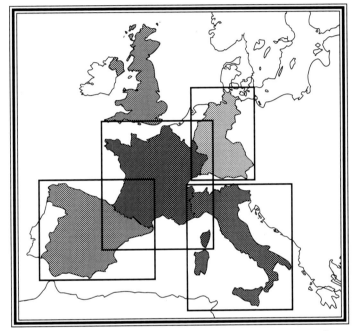

A study of the regulatory arrangements in
France, West Germany, Italy and Spain

HSE
Health & Safety
Executive

© Crown copyright 1991
First published 1991
ISBN 0 11 885614 6

General enquiries regarding this publication should be
addressed to the Health and Safety Executive at one of
the following public enquiry points.

Health and Safety Executive
Library and Information Services
Broad Lane
SHEFFIELD S3 7HQ
Telephone: (0742) 752539
Telex: 54556

Health and Safety Executive
Library and Information Services
Baynards House
1 Chepstow Place
Westbourne Grove
LONDON W2 4TF
Telephone: 071-221 0870
Telex: 25683

CONTENTS

CONTENTS – (continued)

PREFACE

This study describes the health and safety regulatory systems in France, the former West Germany, Italy and Spain, and compares the accident statistics in these four countries with Great Britain.

In February 1990 the Health and Safety Executive began work on a series of studies of approaches to health and safety regulation, including enforcement, in the four other largest member states of the European Community: France, the former West Germany, Italy and Spain. The report draws together descriptions of the main features of the health and safety systems and presents a comparison of fatal and non-fatal accident rates. The report also contains descriptive and statistical details about each country.

In certain aspects the systems in the four continental countries differ from Great Britain: in particular, in none of these countries is there a single organisation with responsibility for overseeing standards of health and safety in the workplace; comprehensive accident insurance arrangements on a no-fault basis are an important feature of all the major continental systems; and in France and Germany the accident insurance organisations in effect provide specialised inspectorates to complement the labour inspectorates, which themselves have responsibilities going much wider than health and safety.

The report shows that British fatal accident rates are substantially lower, both for individual industrial sectors and for all-industries combined, than in France, Italy and Spain (apart from agriculture), and somewhat lower than in West Germany in manufacturing and services. It also suggests that a similar position applies for non-fatal injuries. The report concludes that differences in the pattern of economic development over the post-war period might have a part to play in explaining relative accident performance; but differences in regulatory systems may also be part of the explanation.

It is hoped that this study will encourage policy makers in Europe and elsewhere to undertake further investigation of the mechanics of enforcement of health and safety legislation and to focus on the complexities of the different systems in different countries.

GENERAL OVERVIEW

Introduction

1   This report presents the findings of a study carried out by the Health and Safety Executive (HSE) of the health and safety regulatory systems in France, the former West Germany, Italy and Spain. The study was undertaken (in part at the prompting of the Parliamentary Select Committee on Employment) during the course of 1990 in order to obtain a better appreciation of the legal and institutional framework influencing the way in which other major European Community (EC) countries deal with the problems and challenges of making workplaces safer and more healthy.

2   The report outlines the legal framework in France, the former West Germany, Italy and Spain and examines the countries' institutional arrangements for inspection, occupational medicine arrangements, research and technical support, and for involving the social partners in decisions on health and safety matters. It compares the incidence of industrial accidents in these countries with Great Britain and goes on to provide more detailed information on each continental country. The information has been culled from literature searches and brief visits to the countries concerned.

3   It was not a main purpose of the study explicitly to compare practices in other EC countries with those in Great Britain. We have therefore attempted no description of British institutional arrangements. But in examining the approach in other countries we have highlighted a number of areas where it was of particular interest to find out whether arrangements in Britain were in general broadly similar to, or differed markedly from, those in the four other largest EC member states. For example, do the other countries have bodies like HSC/HSE, tripartite in composition and covering all the main areas of industrial and occupational risks to the health, safety and well-being of people at work? Is there a relatively centralised or decentralised regime for the regulation of these risks? Is greater emphasis placed on certification of plants before they are brought into operation or on regulation/control of processes after installation? And how do other countries tackle the problems of occupational health? While comparisons of this kind will not necessarily indicate how effective other countries are in controlling or reducing the numbers of accidents, they can provide valuable insights into alternative methods which may be of use to policy makers: and this, of course, is ultimately the purpose of the study.

4   In general, the area to which our attention was directed by the way the institutions operate was that covered in Great Britain by the Field Operations Division of HSE (the factory and specialist inspectorates) plus the equivalents of HSE's Research and Laboratory Services Division and Directorate of Information and Advisory Services. Certain of the countries under study had, for example, mining or nuclear inspectorates which did not form a part of the central administration of health and safety as they do in Great Britain; and similarly our study did not cover or identify responsibility for such matters as the transport of dangerous goods, the totality of the controls on hazardous substances, or many other aspects of health and safety protection affecting the public or the wider environment, such as are dealt with in the main by HSE. The comparisons we make are therefore between the equivalents in other countries of part only of HSE's operations.

The legal framework

5   All the countries studied have, like Great Britain, basic enabling laws establishing general principles governing health and safety at the workplace, as well as subsidiary legislation setting out more detailed requirements. However, an important feature which distinguishes the four countries from Britain, is the prominent role played by accident insurance law (and institutions). To get a complete picture of the legal requirements, one needs to look at the general principles both of labour law and of social insurance law. Another important difference, though this is a matter of judgement and general impression rather than based on the existence of specific references, is that in Great Britain the tradition of regulations is rather less concentrated on the assurance of initial integrity of plant, and more highly concentrated on the safety of processes in operation, than in other Western European countries; save in the major hazard and nuclear areas where all the major countries have substantial arrangements for guaranteeing plant integrity or for the submission of safety cases.

6   The accident insurance arrangements of all four continental countries have the following features:

(a) premiums are paid by employers, although the systems are controlled by the social security ministries; in Spain, the administration of the system is devolved to employers' associations who act as managing agents for the state;

(b) compensation is paid regardless of fault; in theory, the accident insurance associations can claim back the compensation paid from employers if they are proved negligent, but in practice, this does not happen;

(c) employers' premiums can be varied according to an assessment of the accident record of the firm and the risk inherent in the industry;

(d) employees can sue employers at civil law only for pain and suffering, not compensation as such; and in practice, since compensation is generous, civil litigation rarely occurs;

(e) accidents at work have to be reported to the insurance associations, not to the inspection authorities.

There are, of course, no comparable systems in the UK, where indeed the last comprehensive measure (the industrial injuries scheme) was abolished in 1983.

7   In West Germany and France, the accident insurance associations are overseen by the social partners. West Germany's accident insurance associations make their own legal regulations on technical matters and have a separate technical inspection force. France's accident insurance associations also have technical advisers who can be considered in practice to be insurance inspectors. Spain and Italy do not have separate insurance inspectorates.

General provisions and legal duties

8   Very broadly speaking, in France, West Germany, Italy and Spain there is a general requirement in labour law for the employer to protect workers' health and safety and there is also a general requirement in social insurance law for the employer to take steps to prevent accidents. In all countries the general duty is qualified in some way (eg with regard to what is the accepted standard, what is technically feasible etc); but general duties are often subordinated to more specific rules about the matters in question. Subsidiary legislation deals with the establishment of appropriate inspectorates, occupational health arrangements, standards, and major hazard precautions. In none of these countries does there appear to exist a general duty to which undertakings have to have regard for the protection of the public akin to Section 3 of the Health and Safety at Work etc Act.

9   Continental countries tend to regulate the installation of highly hazardous plant in a more prescriptive way than Great Britain, in that they have licensing requirements before plants can legally operate and set specific legal requirements to be enforced by inspectors in addition to placing a general duty of care on the employer.

Use of sanctions

10   In all the four continental countries provision is made for criminal sanctions for breaches of safety legislation. These are generally applied against the responsible individual, not against the company as such. It is difficult to find out the extent to which criminal sanctions are actually used; from what we can establish, only the French make great practical use of them (although the Spaniards have a system of administrative fines; see paragraph 14).

11   In West Germany, labour inspectors can issue enforcement and prohibition notices. If an employer fails to comply, the labour inspectorate can impose an administrative fine. In cases of very serious accidents, the inspectorate can refer the case to the public prosecutor for action under criminal law. But the number of cases referred appears to be very small. The technical inspectors of the accident insurance associations have similar powers of enforcement to the state labour inspectors, but these are founded in social law and cases would ultimately go to the social court rather than the administrative court. Like the state inspectors, the technical inspectors can ultimately refer a case to the public prosecutor if they believe a criminal offence has been committed; but this seldom happens. In addition, as already mentioned, individuals can sue at civil law for pain and suffering arising from accidents at work.

12   In France, inspectors can issue enforcement notices on employers to require them to take health and safety measures and can institute criminal proceedings against responsible managers by presenting papers to the public prosecutor.

13   In Italy, accident investigations are undertaken by magistrates, with the assistance of inspectors either from local health units or the labour inspectorate, and with the power to instigate criminal proceedings in serious cases. The inspectors of the local health units also have the power to issue enforcement or prohibition notices. Inspectors can, in addition, go to a magistrate to institute proceedings in the case of an employer's contravention of health and safety law.

14   In Spain, the law makes provision for administrative (not criminal) fines and these are divided into three categories: 'light'; 'serious'; and 'very serious'. These fines apply to contraventions of wages and employment law as well as health and safety legislation. The appeal process is largely administrative (ie to different levels in a ministry of labour), but further appeals can be made to the courts once the administrative route has been exhausted. In cases of serious accidents or fatalities the criminal law can be invoked, but this rarely occurs. Social insurance law is also relevant because, if a work accident is considered by an inspector to be due to a breach of health and safety provisions, the inspector can recommend that the employee is paid increased compensation by the employer. In addition, the judge in criminal cases may determine the extent of private civil liability, for example, for injuries caused.

15   Our assessment overall would be that the French make a rather greater day-to-day use of criminal sanctions than the British, and both make a very much greater use of them than any of the other countries, with the British additionally applying enforcement notices on lines not dissimilar from that of the Germans. This judgement is based on an Organisation for Economic Co-operation and Development (OECD) classification of enforcement actions, as detailed in the health and safety chapter of the *1990 Employment Outlook*, and which indicates that in 1987 Great Britain took legal proceedings in 2812 cases, France in 10 522 cases and West Germany in 177 cases.

16   In addition to these sanctions, the insurance associations have the power to vary the employers' premium for accident insurance (see paragraph 6) to reflect their accident record: this is intended to act as an incentive to improve safety standards as well as to relate premiums to risk. In practice, these variations in premiums are made according to an assessment of the risk inherent in the industry plus the size and accident record of the firm.

Institutional arrangements

Inspection

17   There are some important differences in the inspection arrangements from country to country. West Germany has a dual system of inspection. State labour inspectors are employed by the federal states and technical inspectors are employed by the industrial accident insurance associations. The state labour inspectors deal with environmental matters and working conditions as

well as health and safety; they spend only 10 to 15% of their time on health and safety matters (a situation very similar to that of environmental health officers in British local authorities). The technical inspectors of the accident insurance associations are industry-sector based and have no responsibility for environmental protection or for general working conditions. They spend about 20% of their total time training employers' safety managers etc and on publicity activities, as distinct from carrying out inspection and giving advice.

18   In France, there is a *corps* of labour inspectors employed by the three ministries of transport, agriculture and labour. The labour inspectors deal with general employment and industrial relations matters as well as health and safety and spend only 30% of their time on health and safety issues. In addition, 'advisory engineers' and 'safety controllers', employed by the industrial accident insurance association, visit firms to investigate accidents and provide advice to employers. These advisers have therefore a significant 'inspection' role.

19   There are also other organisations involved in France. An organisation for the prevention of accidents in construction and public works has a number of engineers who visit construction and related activities and provide what amounts to a state-run advisory service, financed by levies on employers. A national agency for the improvement of working conditions is another state-run (and financed) body which is available to employers to provide advice. Finally, mining engineers inspect 'classified premises' which present a major industrial hazard (eg chemical plants), and also a lower tier of industrial undertakings: their main responsibility is to assess environmental risks, and certify plant integrity.

20   In Italy, labour inspection for health and safety purposes has, since 1978, been the responsibility of inspectors of local health units, which come within the jurisdiction of a ministry of health, rather than a ministry of labour. The latter however retain some powers (eg to investigate accidents) and in some parts of the country (where local health units are not functioning fully) they continue to undertake safety inspection work. The labour inspectorate is also responsible for aspects of nuclear safety, radiation protection and enforcement of labour contracts, as well as for a wide range of employment and industrial relations matters. The accident insurance association receives accident reports but does not play an active role in accident prevention.

21   In Spain, a labour inspectorate covers social security and employment law as well as health and safety; about 15% of the inspectorate's time is spent on health and safety. Technical and specialist assistance to labour inspectors is provided by a national institute of health and safety at work, a division of the ministry of labour (ie there is no specialist inspectorate as such). The accident insurance associations and the labour inspectorate receive reports of accidents at work, but investigation and prevention are the job of the labour inspectorate.

22   It can be seen from these descriptions that in other European countries health and safety at work tends to be the business of several inspectorates or *quasi*-inspectorates, often split between two or more government departments, and that the generalist labour inspectors spend a minority of their time on health and safety issues, backed up by specialist experts from a number of quarters. As stated earlier, we did not explore the means by which the different countries deal with nuclear, mining or other matters which seem, in general, to be only loosely connected with other aspects of the regulation of industrial safety and health.

Occupational medicine

23   Continental countries have a long tradition of medical support at the workplace. This doubtless reflects the history of overall health care provision in these countries where there are no institutions fully comparable to Britain's National Health Service and where the workplace has long been a focal point for the surveillance of health.

24   In West Germany, both the federal states and the accident insurance associations provide occupational health services which have qualified specialist doctors. Employers are required to appoint works doctors, except in small firms. Regular medical examinations are statutorily required for a large number of occupational groups; these are conducted by a doctor (usually the works doctor) authorised by the state medical inspector in conjunction with the relevant accident insurance association.

25   In France all companies have to provide medical surveillance for their employees; medical services can be organised either within the company or shared between a number of small and medium sized companies. All employees must be examined periodically and occupational doctors are also required to observe people at their work places to see if there are any potential health problems.

26   In Italy local health units, run by a ministry of health, are responsible generally for the health of the community, including occupational health. Regular medical examination of the workforce is required by a works doctor.

27   By contrast, Spain does not have a separate medical inspectorate. Regular medical examinations of workers are required, but these are carried out either by company doctors or by occupational health specialists provided by the accident insurance associations. Currently, new legislation is being drafted which contemplates an expansion of occupational health services and, in particular, proposes the creation of trained health specialists in firms.

28   Lists of prescribed occupational diseases vary between continental countries. Workers suffering from these diseases are entitled to compensation from the accident insurance fund (which tends to be more generous than sickness compensation). Comparisons of occupational disease rates, both fatal and non-fatal, are much more problematic than comparisons of work accident injuries (on which see paragraphs 34 to 37). There are a number of reasons for this which are discussed in paragraphs 35 to 39 of the comparison of health and safety statistics which follows this overview. Because of this we have not been able to determine how Great Britain's record compares with other EC member states.

29   In short, the continental countries continue to rely apparently quite heavily on medical surveillance at the workplace by appointed factory doctors, whereas in Great Britain we have a relatively small Employment Medical Advisory Service, and otherwise leave workplace health to private initiative or (in effect) to the National Health Service.

Research and technical support

30   The continental countries derive their research and technical support on health and safety from specialist institutes, mostly independent of government direction but funded either by government or the accident insurance associations. West Germany has two of these: one, funded by government, giving technical support and advice to a federal ministry and state labour inspectorate and a second, funded by the accident insurance associations, which offers similar support to a technical inspectorate. The business of both these institutes is to conduct research and give advice both to the government and to industry. In France, there is a national institute for safety

research which is responsible for collecting and disseminating information on technical aspects of occupational health and safety; this is funded by a national accident insurance association. There is also a national institute for the industrial environment and hazards which, *inter alia*, conducts research into a wide range of health and safety problems arising from industrial activities, and is linked to the ministry of environment. In Italy, there is a government-funded research body which deals with occupational health and safety research and technology; and similarly, in Spain, a national institute of health and safety at work (a division of the ministry of labour) provides specialist health and safety advice to the labour inspectorate.

31   It is worth noting that the research institutes and accident insurance associations tend to be the source of most of the guidance and publicity on health and safety matters and to play the major role in safety training. Thus there is, effectively, a clear organisational divide between the regulatory/inspection function, on the one hand, and the research and technological support work on the other – in contrast to the situation in Britain where policy, inspection and research/technological support are brought together within the single integrated structure of HSE.

The role of the social partners

32   The involvement of the social partners is a significant aspect of the institutional arrangements in the four countries, but takes different forms. In Germany, employers and unions jointly manage the accident insurance associations, while in France, Italy, and Spain employers and unions are represented on the governing bodies of their respective research institutes, and on advisory tripartite commissions. In no country, however, is the determination of policy (and the preparation of regulations) at national level the responsibility of a tripartite commission, as is the case in Great Britain.

33   At the level of the firm the unions exercise substantial influence on company policy and practice, because in all four countries there are requirements, enshrined in law or in collective agreements, for the establishment of joint committees to oversee the management of safety at the company or workplace. In France, an important role is given by law to health and safety working conditions' committees. These have a proactive role and can call in inspectors and experts at the company's expense. The German works councils are well-established within the co-determination framework; and in Italy and

Spain safety councils are also a feature of company organisation.

Accident statistics

34   A recent review of industrial accident statistics carried out by the Organisation for Economic Co-operation and Development (OECD) found that reported fatal and non-fatal industrial accident rates in Great Britain throughout the post-war period have been substantially lower than those for other member states of OECD including France, West Germany, Italy and Spain, although since 1975 some of these other countries, in particular France and West Germany, appear to have achieved proportionately greater reductions in accident rates, albeit from a much higher starting point, than Great Britain. The focus of this comparison was upon features and trends common to all major OECD countries and, while OECD drew attention to differences in the bases of national accident statistics used in the comparisons, little attempt was made to standardise accident rates (beyond removing commuting accidents). There are, however, significant differences in the way individual countries define and record work accidents and diseases, which raised doubts about whether British accident rates were really significantly lower than those of the other main OECD countries. For this reason HSE's Statistical Services Unit has undertaken its own analysis of accident statistics from West Germany, France, Italy and Spain, to adjust for the differences between the way these countries and Great Britain define and record work accidents and also to examine how far differences in the composition of employment are responsible for differences in national accident rates.

35   The results of the HSE analysis of accident statistics are described in detail in the reports on each of the four countries. These results, supplemented in some cases by the results of the OECD comparisons, are discussed in the papers which follows this overview. They indicate that British rates of fatal accidents are substantially lower than those of Spain and Italy, both within each industrial sector and for all-industries combined (although the Spanish fatality rates for all employed people in agriculture are much the same as British rates). They are also substantially lower than those in France for all-industries combined and for each individual industrial sector, with the exception of agriculture, where fatality rates are similar. British fatal injury rates are less than those in West Germany in manufacturing, services and in industry as a whole, but are similar in construction. (In

agriculture the comparison is more uncertain and British rates may be slightly higher than West German rates). For non-fatal injury rates, unless under-reporting of accidents in Britain is very much higher than has been estimated, France and West Germany would appear to have higher accident rates than Great Britain. Differences in employment patterns appear to account for only a small part of these differences in all-industry accident rates.

36   Although definitional differences and differential reporting mean that no meaningful comparisons of levels of serious injuries across the countries can be made, such differences have not changed in recent years and therefore comparisons of serious injury trends can be made. Between 1986 and 1988 West German fatal and serious injury rates for all industries fell faster than those of Great Britain. However, Great Britain's position appears to have been better than that of France, where rates have either risen faster (particularly in construction and transport) or not fallen as fast, and Spain, where rates have risen in all main sectors. Short-term increases in serious injury rates appear to be associated with increases in economic activity reflected by increased employment in Great Britain, France and Spain, but not in West Germany where rates have mainly been decreasing despite increases in employment.

37   These results show that Great Britain does have a significantly lower all-industry rate of both fatal and non-fatal work accidents than its major EC partners, although the differences are not as large as OECD comparisons suggest and West Germany appears to be catching up. While these results suggest that British workplaces are safer, they do not explain why. It is reasonable to suppose that differences in patterns of economic development (discussed in the comparison of health and safety statistics) are a factor in this. However, it is also reasonable to suggest that the relative strengths of the British safety system as a whole and the role played by the regulatory authorities play a part in explaining the

differences observed. The narrowing of the gap between Great Britain and West Germany and increases in serious injury rates in some sectors in recent years are indications that Great Britain's relatively favourable safety performance cannot be taken for granted.

Conclusion

38   This report is not designed to compare the merits of different countries' institutions; its purpose is rather to provide a better understanding of the differences in social and legal contexts. Whether they provide in practice a more effective structure for regulating health and safety at work (in influencing the behaviour of employers and workers) compared with the parallel arrangements in Great Britain, is a question beyond the scope of this study. What this report does bring out, however, is the marked difference in the role played by accident insurance arrangements in the four other largest European Community countries. The influence of accident insurance law is evident, most notably, in the existence of dual inspection systems in France and Germany – and to a lesser extent also in Italy. All four countries have, moreover, given considerable emphasis in their respective legal frameworks to the monitoring of occupational health at the workplace. They have also given more prominence than Great Britain to the safety of work equipment, based in some countries (notably Germany) on a very extensive concentration on industrial standards that we did not examine and whose subject matter is beyond the scope of this study. On the other hand, Great Britain has a regulatory structure in which policy requirements, inspection and research appear to be more tightly integrated than the comparable institutions in the other four countries. On the test of accident rates alone, Great Britain compares favourably with the other four countries examined; but it is less clear, and there are no reliable statistics to settle the matter either way, how the comparison would point as regards occupational health.

COMPARISON OF HEALTH AND SAFETY STATISTICS

by Neil Davies, Head of HSE's Economics Advisers Unit, and Graham Stevens, Statistician, HSE's Statistical Services Unit

Introduction

1   This paper reports the results of comparisons of industrial accident and occupational disease statistics for Great Britain with those for West Germany, France, Italy and Spain. It includes the results of a recent review of industrial accident and occupational disease statistics carried out by the Organisation for Economic Co-operation and Development (OECD). This found that fatal and non-fatal industrial accident rates in Great Britain throughout the post-war period have been substantially lower than those for other member states of OECD including France, West Germany, Italy and Spain, although some of these other countries, in particular France and West Germany, appear to have achieved proportionately greater reductions in accident rates since 1975, albeit from a much higher base than Great Britain. The focus of this comparison was upon features and trends common to all major OECD countries and, while OECD drew attention to differences in the bases of national accident statistics used in the comparison, little attempt was made to standardise the accident rates (beyond removing commuting accidents). There are, however, significant differences in the way individual countries define and record work accidents and diseases, which raised doubts about whether British accident rates were really significantly lower than the other main OECD countries. For this reason HSE's Statistical Services Unit undertook its own analysis of accident statistics from France, West Germany, Italy and Spain, to adjust for the differences between the way these countries define and record work accidents and the way Great Britain does this, and also to examine how far differences in the composition of employment are responsible for differences in national accident rates. This has been a complex and difficult task which could only be done for all four countries for fatal accident rates, supplemented, for France and West Germany only, with a comparison of non-fatal accidents. The comparisons have also been limited to recent years. In addition, recent trends in fatal and non-fatal accident rates have been compared for Great Britain, France, West Germany and Spain.

2   The results of this comparison by HSE statisticians, supplemented in the case of Italy by the results of the OECD comparison, indicate that British rates of fatal work injuries are substantially lower than those in Spain and Italy, both within each industrial sector and for all-industries combined (although the Spanish fatality rates for all employed people in agriculture are much the same as British rates). British fatal accident injury rates are also substantially lower

than those in France for all-industries combined and for each individual industrial sector, with the exception of agriculture, where fatality rates are similar. British fatal injury rates are less than those in West Germany in manufacturing, services and in industry as a whole, but are similar in construction. In agriculture the comparison is more uncertain. At worst, rates may be slightly higher in Britain than West Germany. At best, they are much the same. For non-fatal injury rates, unless under-reporting of accidents in Britain is very much higher than has been estimated, France and West Germany would appear to have higher accident rates than Great Britain. Differences in employment patterns appear generally to account for only a small part of these differences in all-industry accident rates. The effect of differences in employment patterns is greatest in the comparison with West Germany where adjustment for such differences would reduce the overall German fatality rate by up to around a quarter.

3   Between 1986 and 1988, West German fatal and serious injury rates for all industries have fallen faster than those of Great Britain. However Great Britain's performance appears to have been better than that of France, where rates have either risen faster (particularly in construction and transport) or not fallen as fast, and Spain, where rates have risen in all main sectors. Short-term increases in serious injury rates appear to be associated with increases in economic activity as reflected in increases in employment in Great Britain, France and Spain for most main industrial sectors with the notable exception of services other than transport. For West Germany however, rates have mainly been decreasing despite strong growth in economic activity and hence in employment, although figures for all-reported injuries suggest a slowdown in the rate of decline in construction which saw particularly rapid growth in activity and employment.

4   These results show that Great Britain does still have a significantly lower all-industry rate of both fatal and non-fatal work accidents than its major EC partners, although the differences are not as large as OECD comparisons suggest and West Germany appears to be catching up. While these results suggest that British workplaces are safer, they do not explain why. It is reasonable to suppose that differences in patterns of economic development (discussed in paragraphs 40 to 47) were a factor in this. However, it is also reasonable to suggest that the relative strengths of the British safety system as a whole and the part played by the regulatory authorities were a significant factor. The increases in serious injury rates in recent years in some sectors in Great Britain and the

narrowing of the gap between Great Britain and West Germany, are indications that Great Britain's relatively favourable position cannot be taken for granted.

Accident statistics

5   OECD carried out a comparison of accident rates in the major OECD countries which was published in their *1989 Employment Outlook*. The focus of this comparison was upon features and trends common to all major OECD countries and, while OECD drew attention to differences in the bases of the national accident statistics used in the comparison, no attempt was made to standardise the accident rates beyond removing commuting accidents.

6   The comparisons which can be made on the basis of the OECD figures are therefore limited. This is because the close relationship of accident statistics to national systems of safety regulation and social security provision results in significant differences in the way individual countries define and record work accidents. In France, West Germany, Italy and Spain the main sources of data about work accidents are insurance and social security claims. In Great Britain the main sources are reports submitted by employers to enforcement authorities. Among the most significant differences that result are the following:

(a) commuting accidents and traffic accidents arising in the course of work duties are covered by insurance arrangements in all the four selected continental European countries, but are not covered in the British reporting regulations and are thus not included in British industrial accident statistics;

(b) all employees and self-employed people (with the exception of some individuals working on their own) are covered by the reporting requirements in Great Britain and all employees and self-employed people are covered by insurance in West Germany. Hence coverage is nearly comprehensive in both countries. In the other countries it is more limited. In Spain and France insurance is limited largely to private sector employees only (though agricultural self-employed workers are also covered in Spain) – amounting to 65% to 70% of all in employment. In Italy most non-manual and many self-employed people (ie those employees and self-employed people not working with office machinery) are not covered by the Istituto Nazionale per

l'Assicurazione contro gli Infortuni sul Lavoro (INAIL) – Institute for National Insurance against Accidents at Work. Hence coverage is largely limited to manual employees;

(c) because of the link to insurance arrangements (which provide death and disability pensions, partial compensation for lost earnings and medical costs) there are very powerful incentives in all the continental European countries for accident victims and their dependents to ensure that injuries are reported. However, because this insurance is funded wholly by employers there are powerful pressures on the insurance institutions to prevent over-reporting, ie false claims. Because Great Britain relies on reports from employers to enforcing authorities, there are no financial incentives to report. British non-fatal accident statistics are known to suffer from substantial under-reporting, especially of non major over-3-day accidents;

(d) fatalities in France and West Germany are counted where death occurs before award of pension. In France this usually involves a period of between 2 to 3 months; in West Germany usually at least 13 weeks. In the UK a fatality is counted as such if it occurs within 12 months of the accident. This may not significantly affect comparisons because most deaths occur soon after the accident;

(e) West Germany, Great Britain and Italy record non-fatal accidents only where over-3-days' absence from work results. In France all non-fatal accidents involving a day's absence are recorded, while in Spain reporting of all accidents which have caused injury, regardless of length of absence from work, is required.

7   These differences raise doubts about whether the British occupational fatality and non-fatality accident rates are really significantly lower than all the other main OECD countries, as suggested by the OECD comparison (see Charts 1 and 2). Moreover the OECD comparison suggests that over the past three decades some of the other OECD countries have achieved sharper reductions in both fatal and non-fatal accident rates than Great Britain. Both France and West Germany appear from the OECD comparison to have achieved very large reductions in their fatal accident rates since the mid 1960s (France from 19.9 per 100 000 employees in 1965, to 7.6 in 1987 and West Germany from 19.2 per 100 000 full-time workers in 1965, to 5.5 in 1986) although both

Chart 1

Occupational fatality rates (a)

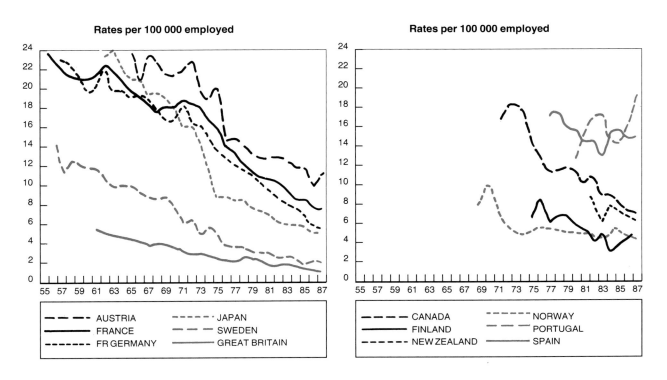

Rates per 100 000 employed

AUSTRIA
FRANCE
FR GERMANY
JAPAN
SWEDEN
GREAT BRITAIN

Rates per 100 000 employed

CANADA
FINLAND
NEW ZEALAND
NORWAY
PORTUGAL
SPAIN

Index numbers (b)

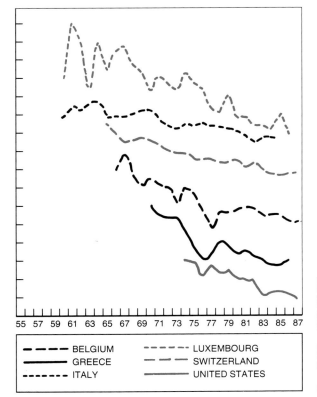

BELGIUM
GREECE
ITALY
LUXEMBOURG
SWITZERLAND
UNITED STATES

(a) Rates are shown as incidence raters per 100 000 employed, with the exception of Belgium, for which frequency rates per 1 million working hours are given. In the main, data refer to employees only, although for Austria, Canada, Germany, New Zealand and for the agricultural sector in Italy, varying proportions of the self-employed are contained in the statistics. In Germany, Switzerland, the United States and for private industry in Italy, rates are calculated on the basis of 100 000 'full-time equivalent' workers.

(b) Each individual on the vertical scale represents a change in the occupational fatality rate of 20 percent. Indices are graphed at different levels for presentational purposes only.

Source: *OECD Employment Outlook 1989*

Chart 2

Occupational injury rates (a)

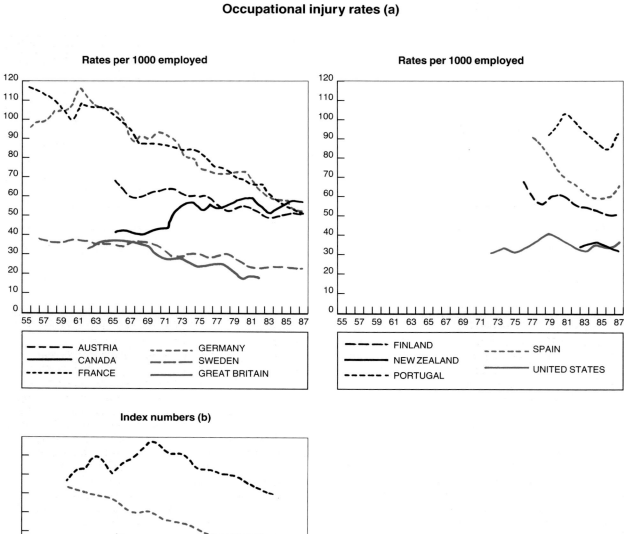

Rates per 1000 employed

AUSTRIA
CANADA
FRANCE
GERMANY
SWEDEN
GREAT BRITAIN

Rates per 1000 employed

FINLAND
NEW ZEALAND
PORTUGAL
SPAIN
UNITED STATES

Index numbers (b)

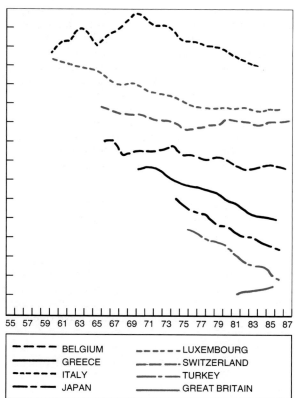

BELGIUM
GREECE
ITALY
JAPAN
LUXEMBOURG
SWITZERLAND
TURKEY
GREAT BRITAIN

(a) Rates are shown as incidence rates per 1000 employed, with the exception of Japan and Belgium for which frequency rates per 1 million working hours are given. In the main, data refer to employees only, although for Austria, Canada, Germany, New Zealand and for the agricultural sector in Italy, varying proportions of the self-employed are contained in the statistics. In Germany, Switzerland, the United States and for private industry in Italy, rates are calculated on the basis of 1000 'full-time equivalent' workers.

(b) Each individual on the vertical scale represents a change in the occupational injury rate of 20 percent. Indices are graphed at different levels for presentational purposes only.

Source: *OECD Employment Outlook 1989*

TABLE 1 Fatal accident rates by industry (a)

	France			West Germany			Italy			Great Britain		
	1965	1975	1987	1965	1975	1986	1965	1975	1984	1965	1975	1985
Total	19.9	15.9	7.6	19.2	13.5	5.5	22.0	20.4	18.2	4.6	2.9	1.9
Agriculture		34.4	12.7	47.1	41.0	17.8	12.1	19.4	16.9		11.7	6.0
Mining and quarrying (b)	60.2	37.4	22.0	6.9	38.8	26.5	42.5	35.3			24.7	20.1
Manufacturing (c)	14.2	10.1	5.9	2.8	10.2	4.1	11.0	7.8	4.4		3.7	2.4
Electricity, gas and water	23.0	13.7	9.6	13.7	10.0	5.7	27.9	12.9				2.0
Construction	51.1	43.5	21.4	30.2	25.7	14.4	81.7	52.7	24.1		17.7	10.5
Trade (d)	11.4	8.8	6.0	9.4	0.7	2.2						0.7
Transport	53.1	50.0	32.7	57.0	32.6	17.1	41.0	33.6	21.4			3.6
Services (e)				3.2	2.2	0.3						0.3

Notes:
(a) Fatal accidents per year 100 000 employees. For Germany and the agricultural sector in Italy, the self-employed are also included
(b) Refers to coal mining for France, Germany and Italy
(c) Refers to metal manufacturing in France, Germany and Italy
(d) Excludes food stores and restaurants in France
(e) For Germany, services refer to health and welfare services only
Source *OECD Employment Outlook 1989*

countries still appear to be markedly worse than Great Britain (whose fatal accident rate has fallen from 4.6 in 1965 to 1.9 in 1985). For Italy the improvement was much less – from 22.0 in 1965 to 18.2 in 1984 (the latest figure OECD could obtain). See Table 1. Much the same trends seem to apply to non-fatal work accidents.

8   Table 1 suggests that fatality rates in Britain are lower for all individual industrial sectors as well as for industry as a whole. However, while comparisons of relative fatality rates by industry reveal similar orderings of industry by risk of fatal accidents in all the countries OECD looked at* there were marked contrasts in the size of the gap between the best and the worst industrial sectors (see Chart 3). Definitional differences may be partly responsible (for example road traffic accidents – included in French accident statistics – these are not evenly distributed across industry). In Great Britain the two worst sectors (mining and quarrying, and construction) were 11 and 6 times worse respectively than the all-industry average for this country, while in West Germany these sectors were only approximately 5 and 3 times worse respectively than its all-industry average. In Spain these sectors were only 3½ and 2½ times worse, and in France only 2 and 3 times worse respectively than the all-industry average. Variations in non-fatal injury rates across industry show similar patterns to these for fatal rates though the size of these variations was less (see

*Mining and quarrying, construction, agriculture and transport tended to have above average fatality rates; other services were below average and manufacturing and utilities were around the average.

Chart 4). Again mining and construction tended to be consistently above average and the service sector, with the exception of transport, consistently below average. Once again Great Britain appeared to have much wider variations across industry than any of the other countries examined, although differential under-reporting of non-fatal accidents may be partly responsible for this. There would appear to be wider differences in safety standards between high and low-risk industries in Great Britain than in the other countries, with possible implications for the development of safety policies. (Targeting attention on high-risk industries would appear to be a more appropriate means of reducing national accident rates in Great Britain than in these other countries). A further implication is that differences in the industrial composition of employment may explain some of the differences between the all-industry accident rates in Great Britain and the other OECD countries.

Comparisons adjusted for differences in reporting arrangements, coverage and employment structure

9   In view of the limitations of the OECD analysis, detailed comparisons of fatal accident statistics from France, West Germany, Spain and Italy and also non-fatal accident statistics from France, West Germany and Spain, have been undertaken by HSE's Statistical Services Unit making adjustments for differences in national reporting arrangements and employment structure. The full results are contained in the individual statistical reports.

Chart 3

Relative fatalilty rates by industry

France
1986

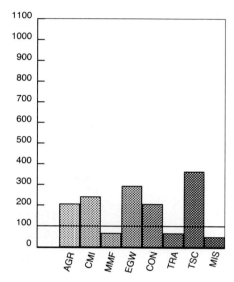

FR Germany
1986

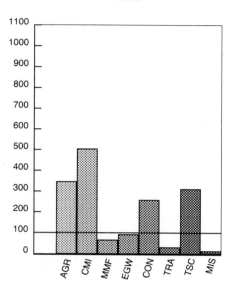

Spain
1987

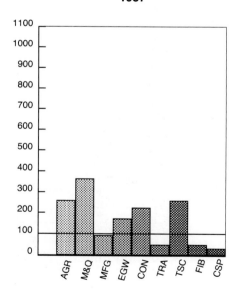

Great Britain
1985

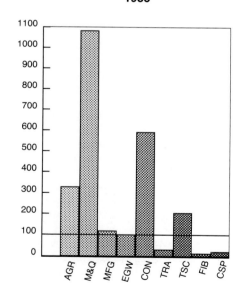

ARG = Agricultural, hunting, forestry and fishing
M&Q = Mining and quarrying
CMI = Coal mining
MFG = Manufacturing
MFF = Metal manufacturing
EGW = Electricity, gas and water
CON = Construction
TRA = Wholesale and retail trade and restaurants
        and hotels
TSC = Transport, storage and communication
FIB = Financing, insurance, real estate
        and business services
CSP = Community, social and personal services
MIS = Miscellaneous services

Source – *OECD Employment Outlook 1989*

(a) Ratios of individual industries to all-industry averages, which are shown as 100. Ratios are given for employees only, with the exception of Germany, for which the self-employed are included in the statistics.

(b) For the majority of countries shown, industries refer to the categories of the International Standard Industrial Classification of Economic Activities (ISIC). In France, hotels and restaurants are excluded from 'wholesale and retail trade', and 'miscellaneous services' refer to the insurance branch 'interprofessional'. In Germany, 'miscellaneous services' refer to health and welfare services only.

Chart 4

Relative injury rates by industry

France
1986

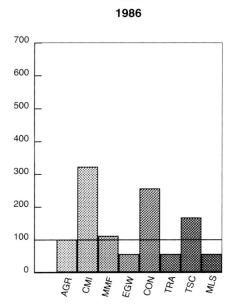

FR Germany
1986

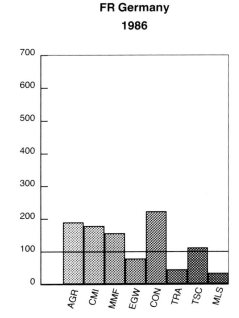

Spain
1987

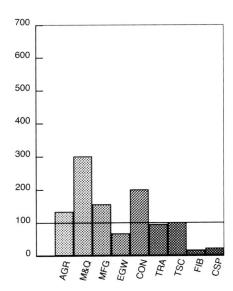

Great Britain
1985

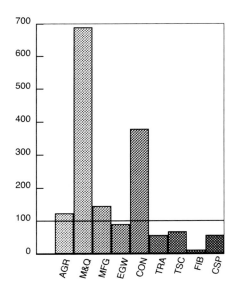

ARG = Agricultural, hunting, forestry and fishing
M&Q = Mining and quarrying
CMI = Coal mining
MFG = Manufacturing
MFF = Metal manufacturing
EGW = Electricity, gas and water
CON = Construction
TRA = Wholesale and retail trade and restaurants
        and hotels
TSC = Transport, storage and communication
FIB = Financing, insurance, real estate
        and business services
CSP = Community, social and personal services
MIS = Miscellaneous services

Source – *OECD Employment Outlook 1989*

(a) Ratios of individual industries to all-industry averages, which are shown as 100. Ratios are given for employees only, with the exception of Germany, for which the self-employed are included in the statistics.

(b) For the majority of countries shown, industries refer to the categories of the International Standard Industrial Classification of Economic Activities (ISIC). In France, hotels and restaurants are excluded from 'wholesale and retail trade', and 'miscellaneous services' refer to the insurance branch 'interprofessional'. In Germany, 'miscellaneous services' refer to health and welfare services only.

TABLE 2 Fatal injuries for all industries except the public sector – standard rates
– rates adjusted to GB industrial mix

Great Britain		Other European Community States						
			France		FR Germany		Spain	Italy
Year	Rate	Year	Standard	Adjusted	Standard	Adjusted	Standard	Standard
1983	2.5	1983	6.3	5.3	5.9	4.7		7.7
1986/87	2.0	1986	4.5	4.1			9.9	(1984) 7.6
1987/88	1.9	1987	4.7	4.2	3.7	2.8	9.7	
1988/89	2.8	1988	5.1		3.7	3.1	10.2	
1989/90	1.9	1989					10.2	
% change		% change						
1983-1987/88	−24%	1983-87	−25%	−21%	−37%	−34%		

Notes:
(a) Rates are per 100 000 employees in GB, Spain and France; employees and self-employed in FRG and Italy
(b) Public administration and education excluded for France, GB and FRG. Some public transport excluded for FRG. Public sector non-manual workers excluded for Spain and Italy
(c) Rates are based on injuries that exclude commuting and work related traffic accidents for Great Britain, France, FR Germany and Spain
(d) Rates for Italy are based on injuries that exclude commuting accidents but include work related road traffic accidents. Exclusion of the latter injuries would be expected to reduce the Italian fatal injury rates by between 30 and 40%. Fatal injuries to some non-manual workers are excluded for all economic sectors

10   Comparisons are easier for fatalities than non-fatal injuries and we are confident that there is no over-reporting in the continental European countries and no under-reporting of fatal accidents in Great Britain. Table 2 shows fatality rates for all industries except the public sector in the latest available years for all five countries. The Table also shows French and West German rates further adjusted to Great Britain's industrial mix. The comparisons show that Great Britain does have a lower all-industry rate of both fatal and non-fatal work accidents than any of the other countries, although the differences are not as large as the OECD work suggests. Table 2 also shows that generally only a small proportion of the differences in all-industry fatal accident rates appear to be due to differences in employment patterns.

11   Of the four selected continental countries, West Germany comes closest to Great Britain. Comparison of fatal accident incidence rates for employees and self-employed people in all industries (excluding public administration and education) in 1987 and 1988 show rates of 3.7 per 100 000 for West Germany in both years, compared with 1.9 in 1987/88 and 2.8 in 1988/89 (which includes fatal injuries from the Piper Alpha disaster) for Great Britain. While the British fatality rate was substantially lower than the West German rate in manufacturing and services there was no significant difference in construction. In agriculture the fatality rate is at best much the same; at worst the rate may be slightly higher in Great Britain (the fatality rate in 1988/89 for employees in agriculture in Great Britain was 7.0 and for employees and self-employed people in agriculture in Great Britain was 8.5, compared to around 7.0 for employees and self-employed people in West Germany in 1987 and 1988). This latter comparison needs to be considered against the much more capital intensive nature of British agriculture which employs only ½ million people (2·2% of total employment), in contrast to West German agriculture, which is still less mechanised and relatively labour intensive, employing 1.3 million people (5% of total employment).

12   How far are the differences in fatal accident rates due to differences in hours worked, or in the pattern of industrial employment? One can assess the significance of possible differences in average hours worked by comparing frequency rates – the number of fatalities expressed per million hours worked. Comparisons of frequency rates for Great Britain and West Germany are given in Table 7 of the statistical paper to the report on West Germany. Comparing frequency rates rather than incidence rates serves to close the gap slightly for all-industries combined, reflecting the greater level of part-time working in Great Britain. However, in some sectors, notably construction, the effect goes the other way. Great Britain had significantly lower fatal frequency rates in construction than West Germany in both years, even though fatal incidence rates were similar, reflecting the significantly greater number of average annual hours worked by construction workers in Great Britain than West Germany.

13   The possible effect of different patterns of industrial employment may be gauged by estimating what the West German fatal accident rate would have been if its distribution of

employment by main industry sector were exactly the same as Great Britains. The effect of this is to lower the estimated fatality rate from 3.7 in 1987 and 1988 to 2.8 in 1987 and 3.1 in 1988. The scale of this change is substantial and reflects the greater proportion of West German manufacturing and service employment in relatively high-risk industries – eg metal manufacture and heavy engineering and transport. A more detailed industrial classification would help explain a little more of the overall difference in fatality rates between the two countries. It is however unlikely that the whole of the difference in fatality rates is due to real differences in patterns of employment.

14   A further effect of employment classifications is the way in which service staff (eg catering and transport workers) are counted in the employing firm's main industry. During the 1980s there was a pronounced tendency for British manufacturing firms to hive off many of these functions to independent contractors whose employment is counted within the service industries to which they belong. We understand that this has not been nearly so much the case in West Germany or elsewhere in Europe. The consequence of the trend towards increased subcontracting in Great Britain is to make the pattern of industrial employment appear more service-sector oriented than it otherwise would, overstating the contrast with the more manufacturing-based pattern of West German employment. Put another way, though marginal, this serves to make British accident rates appear somewhat higher in those manufacturing industries where significant subcontracting has taken place.

15   The comparison of fatality rates between Great Britain and France is more clear cut; the French all-industries rate being more than double the British rate. The French fatality rate is significantly higher than the British rate for each industrial sector, apart from agriculture where the rates are similar. The comparison here for agriculture (as with West Germany) may reflect the much less mechanised and more labour intensive nature of French farming which employs three times as many people as British agriculture.

16   If France's distribution of employment by industry were exactly the same as Great Britains, the overall French fatality rate would be reduced by a further 10% (to 4.1 and 4.2 in 1986 and 1987 compared with actual rates of 4.5 and 4.7 respectively). This reflects the fact that French employment patterns, excepting agriculture, are closer to those in Great Britain than is the case for West German employment patterns.

Standardising the accident rates for both countries on the British industrial pattern of employment widens the gap between French and West German fatal accident rates.

17   The comparison of rates of all reported fatal and non-fatal injuries shows similar rankings, although the size of gap between Great Britain and West Germany in non-fatal accident rates overall and by industrial sector depends on what adjustment is made for under-reporting of over-3-day accidents in Great Britain. The level of under-reporting would have to be over 80% for total accident rates to be the same in both countries. This is very much higher than present estimates of around 50%, derived from comparing accident figures from current reporting arrangements with figures based on previous reporting arrangements (based on claims for industrial injury benefit).

18   If the number of non-fatal injuries per fatality were the same in each country, and hence the comparison of West German and British non-fatal injury rates similar to that for fatal injury rates, then it would suggest a level of under-reporting of British over-3-day accidents of around 65% – or slightly less allowing for the differences in patterns of employment by industry. This is in line with a figure obtained from a pilot of a health and safety supplement to the *1990 Labour Force Survey* and may be confirmed by the results of the Survey itself which will be available during 1991. This would imply that there was some under-reporting of over-3-day accidents under previous reporting arrangements which were derived from individual claims for benefit under the Industrial Injury Benefit Scheme. As mentioned in paragraph 6(c), the incentive to claim and thus to ensure an accident is reported is very strong in West Germany. In Great Britain there is no financial incentive for individuals to ensure an accident is reported under present reporting arrangements and under previous arrangements the incentive was small.

19   The comparison of lost time injuries in Great Britain and France is complicated by the fact that the French count all over-1-day accidents and it is not possible to separate out spells involving 2 or 3-days' absence. This definitional difference is probably not that important. The average duration of lost time injury absences in France is 33 days. There is anecdotal evidence that where British employees are off work due to accidents for more than a day they tend to be away from work for at least four days. The French lost time accident rate is, however, over five times as large as the British and, even allowing for both under-reporting of British accidents and the inclusion of 2 to 3-days'

injuries, would appear not to be lower than the British rate.

20   The comparison for Italy is more limited in that it only covers fatalities in full with little analysis possible of non-fatal injury statistics. It shows that even after making adjustments to remove estimated road accidents and to correct for the omission of many non-manual workers, the fatal accident rate is substantially worse than Great Britains. The estimate of 4.6 for each year 1983 and 1984 appears to put Italy's fatal accident rate slightly below France. This is in marked contrast to the rate quoted for Italy in the OECD comparison of 18.2 in 1984 (Table 1). This is because the figure quoted by OECD is only for insured, mainly manual workers. Since there are relatively few fatalities involving non-manual workers in Great Britain compared to those involving manual workers, an Italian fatality rate, limited largely to manual workers, could be expected to greatly overstate the overall average fatality rate for the employed population as a whole. It is not possible to calculate the overall fatality rate for Italy since the number of deaths to non-manual workers not reported to INAIL is not known. Since there are certain to be some, the estimated fatality rate in the statistical paper to the Italy report understates the true overall fatality rate for Italy, which is probably slightly higher than that for France. Less than 15% of the difference in fatality rates between Great Britain and Italy can be explained by differences in the industry mix of employment.

21   The comparisons of fatality rates for Spain and Great Britain, attached to the report on Spain, indicate that, along with Italy, Spain has a very much higher rate for all-industries combined and individual sectors with the possible exception of agriculture, where rates are similar for all employed workers but, for employees only, British rates are substantially lower. The self-employed account for over two-thirds of Spanish agriculture employment compared to around half of British agricultural employment. This difference, and the much more highly mechanised and capital intensive nature of agriculture in Britain than in Spain, needs to be considered when interpreting the comparison of agricultural accident rates. Only 9% of the difference between Spanish and British all-industry fatality rates can be explained by differences in the industry mix.

Comparison of recent accident trends

22   Fatal occupational injuries are fortunately rare events in all these countries. Consequently comparisons of short-term trends in fatalities can be distorted by the occurence of a single serious accident. Comparison of trends in the broadest category of reported injuries (over-3-day accidents in Germany and Great Britain and over-1-day accidents in France) presents a different problem. The scale of under-reporting of such accidents in Great Britain means that comparison of trends between Great Britain and other countries where reporting is probably complete may be distorted by changes in the level of reporting in Great Britain. Comparisons of recent trends between Great Britain and these other countries (Spain, France and West Germany – relevant figures from Italy are not available) therefore concentrate upon serious injuries, ie fatal and major injuries combined (this is to conform with the presentation by other European countries which publish figures for all injuries, some of which are serious, and serious injuries, some of which are fatal).

TABLE 3 Serious injury rates: all industries except the public sector

Year	GB	France (a)	Germany (a)	Spain (a) (b)
1983		606	191	
1986	100	467	156	132 137
1987	97	449	149	137 142
1988	96	466	139	151 164
1989	95			150p
Change	%	%	%	% %
1983-88		−23	−27	
1986-88	−4	No change	−11	+14 +20
1987-88	−1	+ 4	− 7	+10 +15

Notes:
Rates are expressed per 100 000 employees in GB, France and Spain and per 100 000 insured workers in Germany
(a) Rates exclude vehicle accidents. The rate for Germany in 1986 is based on the estimated number of non-vehicle accidents
(b) Rates include vehicle accidents
p = provisional
Rates for 1986 onwards in GB are for the planning year commencing 1 April. No figure is available for 1983 for GB because the definition of a 'major' injury under the reporting arrangements then in force is not consistent with the definition of a major injury under the reporting arrangements which apply from 1986/87

23   Different criteria are used to define serious injuries and hence no meaningful comparisons of levels of serious injuries between the individual countries can be made but, since these criteria have not changed over the recent period, comparisons of trends can be made. Definitions of serious injuries are given in the statistical papers to the individual country reports. There is some tentative evidence to suggest that West Germany and Great Britain have narrower definitions of serious injuries than France and Spain. For Great Britain serious injuries are taken to comprise fatalities and major injuries, as defined in the reporting regulations introduced in 1986. These

major injuries are mainly fractures (excluding those in the bones of fingers or toes), amputations, burns and other injuries that lead to a stay in hospital for immediate medical treatment for over 24 hours. Rates of serious injury for all industries, excluding the public sector, are given in Table 3 for the years 1983, and 1986 to 1988.

24   In Great Britain the fatal injury rate has been on a downward trend over many decades. Since 1975 it has fallen from 2.8 per 100 000 employees (this is excluding the public sector for consistency with available figures for France, Germany and Spain) to 2.5 in 1983 and 1.9 in 1989/90. Major injury rates rose during the early 1980s, then reached a plateau and, since 1986/87, have fallen back. Over the period 1986/87 to 1988/89 the rate of serious injury (fatal and major) has fallen by 4%. A decline in the rate of serious injury is found in most main sectors of industry. A particular exception, however, is the construction industry where the serious injury rate dropped slightly between 1986/87 and 1987/88 but between 1986/87 and 1989/90 has risen by 5%. Also, the rate of serious injury in transport has risen by 8% between 1986/87 and 1988/89. Rates of all-reported injury have dropped marginally but these are dominated by over-3-day injuries which are subject to substantial under-reporting: there appears to be an upward trend in the rage of over-3-day injuries in the agriculture and manufacturing sectors, but the decline in rates of fatal and major injuries for these sectors suggests that this trend is probably due to somewhat better reporting of over-3-day injuries over the period. The rate of construction over-3-day injuries has dropped by over 2% each year between 1986/87 and 1988/89, but increased marginally in 1989/90.

25   In France the trend of fatality rates, with occasional increases, has been downwards over several decades. Over the past decade the rate for all industries (except the public sector), excluding vehicle accidents, fell from 8.1 per 100 000 employees in 1978 to 6.3 in 1983 and 4.5 in 1986. However, in 1987, the rate went up slightly to 4.7 and in 1988 rose further to 5.1. Particularly sharp increases between 1986 and 1988 occurred in transport (up 27%) and in construction (up 44%). There was also a rise in agriculture of 5% between 1987 and 1988. The rate of serious (permanent incapacity) injuries also rose in 1988 in all main sectors except agriculture, punctuating a steady decline since 1983. In transport the rate increased by 9% and in construction rose by 5% between 1987 and 1988.

26   In West Germany for both fatal and serious (ie fatal and first-time compensated) injuries,

rates have declined in all main sectors throughout the 1980s – although a pit disaster contributed to a near doubling of the fatal rate in the energy sector between 1987 and 1988 (in Great Britain this sector suffered the Piper Alpha disaster). Between 1986 and 1988 rates of serious injury (excluding vehicle accidents) fell by 11% in all industries and by 14% in construction. The rate for transport rose marginally by 2% but examination of published rates (which include road vehicle accidents) per full-time worker suggest this may just be a blip in the long-term decline. The all-industry rate of all-reported (over-3-day) injuries also continued to decline, whether expressed per capita or per full-time equivalent worker. This is reflected in declining rates in all main sectors, although in construction this decline seems almost to have stopped. The general decline in injury rates between 1986 and 1988 comes at a time when employment has risen in all sectors except agriculture and energy. There does however, appear to have been a slowdown in the rate of decline compared to earlier years.

27   In Spain road vehicle accidents can be excluded from all-industry figures but not from figures for individual industrial sectors. This does not, however, appear to affect the trends over recent years. Rates of both fatal and serious injury have been rising during the latter 1980s in all main sectors except for agriculture.

TABLE 4 Fatal injury rates: construction

Year	GB	France (a) (b)		Germany (a) (b)		Spain
1983	11.6	23.0	29.3	13.6		
1986	10.2	16.5	21.4		14.0	35.1
1987	10.3	17.5	21.2	10.2	15.6	32.3
1988	9.9	23.8	28.7	8.9	13.7	39.0
1989	9.4				13.9	36.5p
Change	%	%	%	%	%	%
1986-88	− 3	+44	+34		− 1	+11
1983-88	−15	+ 3	− 2	−35		

Notes:
GB
Rates per 100 000 employees, exclude vehicle accidents occurring on the road. For the years 1986 onward, rates are for the planning year 1 April to 31 March
p = provisional
France
Rates per 100 000 employees. The two figures given:
(a) exclude; and
(b) include, vehicle accidents on the road occurring in the course of work
Germany
(a) Rates per 100 000 insured workers. Vehicle accidents on the road
    occurring in the course of work are excluded
(b) Rates per 100 000 full-time workers. Vehicle accidents are included
Spain
Rates per 100 000 insured employees. Vehicle accidents are included

28 Comparisons of fatal injury rates (excluding road vehicle accidents) show that West Germany achieved a greater proportionate fall, of 37%, between 1983 and 1987, than Great Britain and France where the fall was around 25% in each country. The more recent trends are affected by disasters in the energy sector in both West Germany and Great Britain in 1988. Without these, the rates would have continued to fall in both countries. In France and Spain, in contrast, fatality rates rose by 13% and 22% respectively between 1986 and 1988 (see Table 2).

29 In construction, between 1986 and 1988 the fatality rate (excluding vehicle accidents) fell by 3% in Great Britain and 1% in West Germany but rose by 11% in Spain and 44% in France (see Table 4). It should be noted that the reductions both over the longer period, 1983-88, and the latest year, 1987 to 1988, appear greater in West Germany than Great Britain.

30 Table 3 gives details of the comparative trends in the all-industry rates of serious injuries since 1986. The largest proportionate decline between 1986 and 1988 was achieved by West Germany (a drop of 11%), followed by Great Britain (a 4% fall). Despite a dip in 1987, the French serious injury rate was largely unchanged over the period, while there was a continuous rise in the serious injury rate in Spain (up 14% over the period).

TABLE 5 Serious injury rates: construction

Year	GB	France		W Germany		Spain
		(a)	(b)	(a)	(b)	(b)
1983		1697	1778	353		
1986	293	1355	1414	320	382	294
1987	287	1318	1376	266	365	311
1988	296	1388	1449	275	371	334
1989	308				351	311p
Change	%	%	%	%	%	%
1986-88	+ 1	+ 2	+2.5	−14	− 3	+14
1983-88		−18	−19	−22		

Notes:
GB – Fatal and major
France – Permanent incapacity
Germany – First time compensated
Spain – 'Graves'
p = provisional
All rates per 100 000 employees GB, France and Spain; per 100 000 insured workers in Germany
(a) rates exclude vehicle accidents, the figure for Germany 1986 is based on the estimated number of non-vehicle accidents
(b) rates include vehicle accidents, rate for Germany are for full-time equivalent insured workers
No figure is available for 1983 for GB because the definition of a 'major' injury under the reporting arrangements then in force is not consistent with the definition of a major injury under the reporting arrangements which apply from 1986/87

TABLE 6 Serious injury rates: transport

Year	GB	France	West Germany	
	(a)	(a)	(a)	(b)
1983		591	213	
1986	92.1	511	182	231
1987	96.3	509	199	249
1988	99.1	553	186	235
1989	100.3			220
Change	%	%	%	%
1986-88	+ 8	+ 8	+ 2	+ 2
1983-88		− 6	−13	
1987-88	+3	+ 9	− 7	− 6

Notes:
Rates are expressed per 100 000 employees for GB and Spain and per 100 000 insured workers in West Germany
(a) Rates exclude vehicle accidents. The figure for Germany in 1986 is based on the estimated number of non-vehicle accidents
(b) Rates include vehicle accidents. Rates for Germany are expessed per 100 000 full-time insured workers
No figure is available for 1983 for GB because the definition of a 'major' injury under the reporting arrangements then in force is not consistent with the definition of a major injury under the reporting arrangements which apply from 1986/87.

31 Comparisons of serious injury trends for the construction and transport industries (Tables 5 and 6) also show more rapid improvements in West Germany than the other countries. In construction, serious injury rates dropped in West Germany between 1986 and 1988 (by 14%, excluding vehicle accidents). There was a small rise between 1987 and 1988 but the published 1989 figure (which includes vehicle accidents) shows a continuation of the decline in serious injury rates in West German construction. In contrast in Great Britain, serious injury rates for construction have risen each year between 1986/87 and 1989/90 (a rise of 5% over the period). In France, too, the serious injury rate for construction has increased by 2%, between 1986 and 1988, after a steady fall in earlier years.

32 In the transport industry, despite temporary blips in 1985 and 1988, the serious injury rate in West Germany shows a continuing long-term decline which, from the published rates for 1989, appears to be continuing. Between 1986 and 1988 in West Germany the rate fell by 2% while in Great Britain it rose by 8% and in France it also rose by 8%.

33 In the manufacturing sector the serious injury rate fell by 1% in Great Britain between 1986 and 1988 while the rate fell by slightly more in West Germany. In France it was broadly unchanged while in Spain it rose by 20% (this included road vehicle accidents).

34   For all four countries the period of 1986 to 1988 was one of rising economic activity and employment in most main sectors, particularly in services but also in construction and manufacturing, though to a lesser extent in Great Britain and France (where manufacturing employment fell slightly). Employment in agriculture fell in three out of the four countries (by 5% in Great Britain, 3% in West Germany and 1% in Spain) and was unchanged in France. Employment in the energy sector also fell in Great Britain, France and West Germany (by 9%, 5% and 5% respectively). There is a clear tendency for the trends in serious injury rates to be positively influenced by the trend in economic activity as reflected by the employment growth. The strength of this relationship however, varies between countries and industrial sectors. West Germany appears to be a particular exception; its serious injury rates have fallen in all main sectors, even those where employment has risen sharply, although in construction, which has had the greatest rise in employment, the decline in the rate of serious injury has been slowing and the rate for all-reported injuries has been nearly level. In Great Britain and France serious injury rates in other (non-transport) services have fallen slightly despite strong employment growth. Elsewhere, serious injury rates have risen where employment has risen, while rates have remained steady or fallen where employment has fallen or stayed steady.

Occupational disease

35   Comparisons of occupational disease rates, both fatal and non-fatal, are very much more problematic than comparisons of work accident injuries (and, for the reasons described above, these are difficult enough). The main problems are:

(a) many occupational diseases are clinically indistinguishable from general chronic type diseases resulting from other factors;

(b) epidemiological evidence that a particular condition is related to occupational exposure may take many years to establish, particularly where there is a long latency period;

(c) as a result, the definition of particular diseases as 'occupational' largely reflects a value judgement as to those conditions for which the workplace is to be held responsible and the victims eligible for special compensation. The Spanish list of prescribed occupational diseases contains around 40

Chart 5

Occupational fatality rates

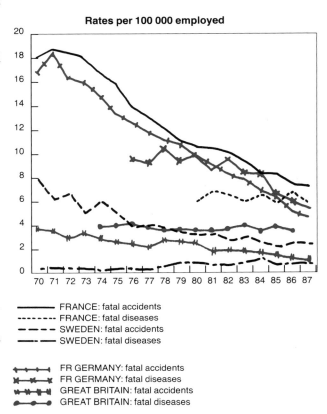

Rates per 100 000 employed

———— FRANCE: fatal accidents
------------ FRANCE: fatal diseases
– – – SWEDEN: fatal accidents
—·—· SWEDEN: fatal diseases

+—+—+ FR GERMANY: fatal accidents
×—×—× FR GERMANY: fatal diseases
■—■—■ GREAT BRITAIN: fatal accidents
●—●—● GREAT BRITAIN: fatal diseases

Germany and Great Britain data refers to all civilian employment: for Sweden data refers to employees, for France to employees in private industry only. In Germany, rates are calculated on the basis of 'full-time equivalent' workers.

Source – *OECD Employment Outlook 1990*

diseases while the French contains more than twice that number;

(d) different national practices regarding health monitoring of workers will influence the likelihood of recognising and recording a condition as occupationally related.

36   Notwithstanding these limitations, OECD carried out a comparison of fatalities from occupational diseases for their *1990 Employment Outlook*. Chart 5 shows occupational fatalities both from diseases and accidents since the 1970s for Great Britain, West Germany, France and also Sweden (insufficient information was available for Spain and Italy). The rates for occupational disease fatalities include those of recipients of disability pensions. The comparison suggests that

the occupational disease fatality rate in Great Britain, while significantly higher than Sweden, continues to be well below that in France and West Germany. While the rate has decreased considerably in West Germany it has remained by and large level in France and Great Britain and increased in Sweden. This is in marked contrast to the rate for fatal accidents which has fallen in all these countries.

37   OECD suggests that, while changes in reporting accuracy and classification systems play a larger role in the area of illness as compared to occupational injury, it is not likely – with the possible exception of Sweden – that they represent an important explanation for the diverging trends. Neither does the explanation appear to lie with additions to national lists of prescribed diseases. Deaths due to diseases added to the lists since the 1960s constitute only a small proportion of the totals.

38   OECD does, however, note that the trends largely reflect levels of exposure to hazards 10, 20 or 30 years earlier. Furthermore, for all the countries OECD looked at (including the four mentioned above, plus others where only more recent data was available on a consistent basis), the figures for deaths due to occupational disease were dominated by deaths of miners from pneumoconiosis. The relative position of different countries in respect of their reported rates of fatalities from occupational diseases are thus to a large extent a reflection of the share of overall employment engaged in mining 20 to 30 years ago and the conditions in that industry at that time.

39   HSE has not attempted to undertake any analysis of comparative occupational ill health statistics for this project because of the difficulties described in paragraph 35.

Social and economic influences upon accident rates

40   In the following paragraphs a number of factors which may affect the relative safety performance of the different countries studied are considered.

41   The OECD comparison of occupational accident trends noted the high plateaus of occupational fatality and injury rates, and overall numbers, reached in the mid 1950s to early 1960s in a number of the major OECD countries examined. It suggested that this reflected the post-war rebuilding of the productive apparatus, rapid economic growth and increasing mechanisation in the countries concerned.

42   The comparisons of the most recent figures suggest that differences in industrial structure explain only a small fraction of the differences in current all-industry fatal accident rates. However, as reported in the statistical annex to the *Health and Safety Commission 1989/90 Annual Report*, approximately a third of the reduction in annual numbers of fatal injuries in Great Britain during the 1980s can be explained by shifts in employment between the main industrial sectors.

43   Various studies have drawn attention to the adverse effect of rapid economic growth upon safety performance, which appears to be reflected in the adverse movement of serious injury rates in some sectors in Great Britain, France and Spain over recent years. A study of the effectiveness of HSE's field activities* found that as indicators of the level of economic activity, increased levels of overtime and the engagement of new workers, both had a significant effect in raising accident rates in Great Britain. A study of changes in accidents at work in Spain between 1977 and 1987 by the Spanish Ministry of Labour and Social Security† also drew attention to the importance of economic factors in explaining the dramatic decrease in Spanish accident rates between 1977 and 1984 and subsequent increase after 1985. The period of falling accident rates in Spain coincided almost exactly with the period of major output declines. The explanation suggested for this was that the economic crisis affected workplaces with older installations and obsolete processes most. Such plants tend to have the worst working conditions and safety. Furthermore, during periods of recession, companies usually function without making full use of capacity, resulting in less crowding and less pressure on the pace of work, all tending to lower accident rates. After 1985 the Spanish economy began a period of strong growth in economic activity and its effect upon accident rates seems to have been almost immediate, with rates in 1987 returning to the levels which existed in 1982/83.

44   While rapid economic growth may help to explain the rises in accident rates observed in the fast growing OECD countries (ie Japan, West Germany, France and Italy) in the early part of the post-war period, and the increasing rates of serious injury observed in the last few years in some sectors in Great Britain, France and Spain, it is less obvious how this explains the declining

Measuring the effectiveness of HSE's field activities HSE Occasional Paper OP11, HMSO 1985, ISBN 0 11 883842 3

†*Accidents at work 1977-87* Spanish Ministry of Labour and Social Security No 26 May 1988

trends over much of the past three decades or the comparative differences which still exist. Throughout most of the 1960s and 1970s the continental Western European countries all experienced higher rates of economic growth, yet France and West Germany have also achieved more rapid reductions in accident rates than Great Britain, although from a much higher level, while Italy and Spain have not. From 1981 through to 1988, Great Britain has achieved somewhat faster economic growth than the rest of Western Europe, and faster than it managed to achieve over most of the preceding post-war period, although it appears to have continued to achieve lower accident rates than its major European Community partners.

45   Part of the explanation may lie in the nature of the post-war social and economic development of these five countries. Great Britain began the post-war period having suffered less war damage and with a substantially more highly developed economy than the other major Western European countries. In contrast to some of the other countries examined, Great Britain also had a long established and widely respected health and safety regulatory regime which was subject to comprehensive reform and modification in the early 1970s. In France and West Germany agriculture accounted for over a fifth and in Italy and Spain over a third of total employment at the start of the 1950s. Both France and West Germany achieved high rates of growth throughout the 1950s and 1960s with industry serving as the locomotive for this growth which was facilitated by mechanisation within agriculture. This released surplus labour which was absorbed into the expanding industrial and service sectors – a pattern of development Great Britain had experienced, at a slower pace, in the nineteenth century and which Italy and Spain experienced only later (Italy from the 1950s, Spain beginning in the early 1960s). Hence, while manufacturing employment began to decline in relative terms in Great Britain as early as 1961 and in absolute terms in 1966, in West Germany and France manufacturing employment only began to plateau off at the very end of the 1960s and has since accounted for a fairly constant share of total employment. In Italy and Spain manufacturing continued to account for a growing share of employment until the middle and end of the 1970s respectively. The large 'informal sector' (or 'black economy') might be a further reason for these countries' higher accident rates: through undercutting firms in the formal sector and forcing employers there to neglect safety to remain competitive.

46   The stage of development characterised by 'extensive' growth of the industrial sector and substantial shifts of employment out of agriculture has a number of features unlikely to be conducive to safety. In particular:

(a) increasing mechanisation within agriculture with a workforce inexperienced in the safe use of such equipment;

(b) a relatively large proportion of the workforce inexperienced and untrained in safe systems of work;

(c) priorities of management within the industrial sector more likely to be focussed on production rather than safety, with many managers themselves relatively inexperienced;

(d) regulatory authorities likely to be unfamiliar with the emerging hazards, and to be inexperienced and overstretched.

47   A more 'mature' industrial structure characterised by growing emphasis upon the service sector and 'intensive' rather than 'extensive' growth of the industrial sector (ie growth achieved more by additional or more advanced capital being applied to the existing workforce rather than increases in all factors of production – land, labour and capital) is likely to be more conducive to safety. In particular, because:

(a) a larger proportion of the industrial workforce and managers will be experienced;

(b) there will be an increased emphasis upon training;

(c) individual workers become more difficult and expensive to replace if injured, encouraging managers to devote more attention to safety in the interests of production;

(d) there is likely to be increasing social and political emphasis upon safety as real incomes rise;

(e) regulatory authorities are likely to be more experienced and become relatively better resourced.

Conclusion

48   The comparison of accident statistics for the four selected continental European countries with Great Britain indicates that, even after allowing for different levels of reporting and excluding those accidents not counted as 'work accidents' in Great Britain, British accident rates are substantially lower overall than those in Spain (except for agriculture) and Italy. They are also significantly lower than those in France (except for agriculture) and significantly lower than those in West Germany in both the manufacturing and service sectors.

49   Between 1986 and 1988 West German fatal and serious injury rates have fallen faster than those of Great Britain although Great Britain's position appears to have been better than that of France, where rates have either risen (particularly in construction and transport) or not fallen as fast, and Spain, where rates have risen in all main sectors. Short-term increases in serious injury rates appear to be associated with increases in economic activity reflected in rising employment in Great Britain, France and Spain for most main industrial sectors, except for non-transport services. For West Germany, however, rates have mainly been decreasing despite an increase in employment.

50   A number of factors would appear to lie behind Great Britain's lower accident rates. The relative strengths of the British safety system as a whole, and the part played by the regulatory authorities within it, though clearly important, may not fully explain this. They are in themselves a part of a complex set of wider social and economic influences which this paper has only touched upon. The increase in the serious injury rate in some sectors in Great Britain in line with increased employment, possibly reflecting the impact of an increased pace of economic activity upon the health and safety resources and commitment of firms, is an indication that Great Britain's relatively good industrial accident record cannot be taken for granted.

REPORT ON THE SYSTEM FOR THE HEALTH AND SAFETY PROTECTION OF WORKERS IN FRANCE

by Lindsay Jackson and Kevin Myers,
Health & Safety Executive

Introduction

1   The first French legislation relating to the protection of workers dates back to 1841. This was concerned initially with the employment of young people. The legislation followed a similar development to that in Great Britain evolving piecemeal over time and gradually incorporating provisions relating to health and safety.

2   It became apparent that the inspection and enforcement of the law was ineffective. This led to an act in 1892 which set up a body of officials whose sole function was factory inspection. The act also provided for the appointment of women factory inspectors. The number of inspectors was increased every time new legislation was passed, making the inspectors responsible for the enforcement of further measures of labour protection.

3   In 1930, a new law was enacted which required preventive measures to be taken by employers for the protection of employees. In 1939, legislation imposed responsibilities on the suppliers and manufacturers of machinery to build health and safety into the design of machines. Further key legislation in 1947 established the state controlled insurance system, set up the National Institute for Safety Research – Institut National de Recherche et de Sécurité (INRS) and required the involvement of workers' representatives in health and safety issues. In 1982, key legislation was introduced concerning workplace committees which considered, amongst other things, health and safety. This included a statutory requirement for training safety representatives. A national commission Conseil Supérieur de la Prévention des Risques Professionnels (CSPRP) – Higher Council for the Prevention of Occupational Hazards – was established.

4   The current health and safety law is primarily contained in the Labour Code and the Social Security Code.

Legal and institutional framework

The Labour Code

5   The Labour Code is administered by the Ministry of Labour and is an amalgamation of major statutes and measures based on those statutes. The first was produced in 1913. The provisions of the Code apply to all industrial, commercial and agricultural establishments, whatever the form of their ownership or management, including family enterprises and government offices, the professions, hospitals and technical training and educational establishments.

6   Mines, quarries and transport enterprises are not covered by the provisions of the Code unless specifically included. Schools and educational establishments not involved in technical training are not covered by the Code. The Code does not cover self-employed people except in very limited circumstances.

7   The Labour Code includes general principles and specific requirements on:

(a) general health and safety responsibilities:

   (i) employers must maintain their establishments in a state of hygiene and cleanliness necessary for the health of their employees, and manage establishments so that the safety of the workers is guaranteed; and ensure that equipment is installed and maintained in the safest way possible;

   (ii) manufacturers, suppliers and importers are prohibited from providing apparatus which is not made or controlled in a way which ensures the health and safety of workers.

(b) safety training:

   (i) the heads of establishments must organise safety training for all types of new and returning workers. The workforce must be consulted on the design and implementation of the training. The organisation and finance of training is also controlled.

(c) health and safety committees:

   (i) committees must be established in firms with more than 50 employees (300 in construction). The membership and function of the committee is set down (see paragraph 10).

(d) occupational health service provisions:

   (i) all employers covered by the Code (including transport) must provide access to occupational health services. All employees must have access to a doctor for medical examination at least once a year. These doctors either sell their services to the company or may be employed by large companies. They only carry out preliminary examinations, any problems identified are referred to the employee's doctor. They advise employers if a pattern of ill health is identified. In addition, occupational doctors spend around one third of their time observing the workers at work so that they can pick up on any health risks as a result of systems or methods of work.

(e) use of dangerous substances and preparations:

   (i) prohibitions and restrictions on the use of harmful agents;

   (ii) packaging and labelling requirements.

(f) other health protection measures:

   (i) restrictions on the exposure of women and young people to harmful agents and occupations;

   (ii) use of primary engineering controls;

   (iii) use of personal protective clothing and equipment in designated processes;

   (iv) provision of information to employees by means of warning notices;

   (v) provision of washing and sanitary facilities;

   (vi) establishment of first-aid and emergency procedures.

(g) safety measures:

   (i) detailed safety measures are laid down for particular processes, activities or industries.

(h) special provisions concerning the construction industry:

   (i) a safety plan must be drawn up for each phase of a construction project.

8   The Code also requires co-ordination between organisations established under the Labour Code and the Social Security Code and covers the enforcement powers of labour and medical labour inspectors.

9   The Labour Code states that heads of businesses, as individuals, are responsible for the health and safety of their employees. In tandem with this responsibility is a favourable attitude of the judiciary towards employers who wish to discipline employees who consistently fail to follow health and safety instructions. It is possible for employers to delegate power to supervising employees to do something in terms of health and safety, but they can never delegate their responsibilities.

10   A health and safety committee Comité d'Hygiène, de Sécurité et des Conditions de Travail (CHSCT) is compulsory for all businesses with more than 50 employees (300 in construction). The chairman of the committee is the employer. The secretary of the committee used to be a representative of the employer but this has recently changed and now the secretary is elected by members of the committee. The employees' representatives on the committee are elected by the workforce. The employer makes final decisions on matters of health and safety but the committee has an important advisory role. The employer is required to pay for at least one week of safety training for at least one member of the committee per year. The employer is required to inform and consult the committee at least once a year about the plan-of-work for the next year and the results of the work undertaken in the previous year. The committee meets once a quarter and has the right to be consulted by the management on health and safety in the company, call in inspectors, and to call in other experts (at the employer's expense) if they require further information or lack the technical expertise to analyse the risks at their place of work.

11   The Labour Code does not deal exclusively with health and safety – it also covers other conditions of employment such as hours of work, contracts of employment, collective agreements and wages.

The Social Security Code

12   The Social Security Code is administered by the Ministry of Health and Social Security and supplements the Labour Code on health and safety matters in the following ways:

   (a) all employers must make contributions to the appropriate sickness insurance fund;

   (b) it specifies the roles and organisation of the National and Regional Technical Committees (paragraph 42), the National Sickness Insurance Fund (paragraph 29) and the Regional Sickness Insurance Funds (paragraph 32);

(c) provisions are made for recognition and compensating occupational diseases and accidents.

13   The Code applies to all employers in industry and commerce, including employers of home workers. The compensation provisions apply to all workers in these sectors and also to unpaid workers (eg charities, prison-labour). Separate similar insurance schemes cover employees in agriculture, mining, the civil service, shipping, fishing and defence.

14   The Code requires tables to be drawn up listing occupational diseases and the circumstances which give rise to them. These form the basis for compensation payments by the Primary Sickness Insurance Funds.

15   On accident and disease prevention, the Regional Sickness Insurance Funds (paragraph 32) are staffed by advisory engineers and safety controllers who are responsible for conducting research on health and safety questions and for visiting workplaces to inspect health and safety provisions.

16   The Code allows the Regional Sickness Insurance Funds to draw up recommendations appropriate to particular activities. These recommendations, which include advisory atmospheric limit values, represent current state-of-the-art as far as health and safety precautions are concerned.

The Public Health Code

17   The Public Health Code is administered by the Ministry of Health and Social Security and covers a wide range of public health issues including the manufacture, import and export, storage trade and use of toxic and dangerous substances. These provisions are also relevant to occupational health matters.

The law on the control of chemical products

18   The law on the control of chemical products is administered by the Ministry of the Environment and is designed to protect people and the environment from hazards that can arise from chemical substances and preparations. It establishes pre-manufacture declarations for a wide range of chemicals newly introduced into the French market and allows restrictions to be placed on product marketing.

The Labour Inspectorate

19   The bulk of labour inspection at the workplace is undertaken by generalist labour inspectors, employed by the Ministry of Labour. There is a common core of inspectors serving the transport, agriculture and labour ministries who all receive the same two year training. The labour inspectors enforce the Labour Code in commerce, industry, hospitals, transport and agriculture. They enforce all parts of the Labour Code including conditions of work, industrial relations agreements and practices and redundancy as well as health and safety – which amounts to about 30% of their activities.

20   The Labour Inspectorate of the Ministry of Labour is organised on a geographical basis into 440 sections, spread across 100 departments and 23 regions (including overseas areas). The Labour Inspectorate in the areas of transport and agriculture is organised in a similar way, although on a much smaller scale. For example, for transport there is often only one inspector for two geographical departments. The 440 sections are headed by an equal number of inspectors, who are assisted by about 900 labour controllers. Each region, in addition, has a medical labour inspector and an engineer to provide advice.

21   Inspectors and controllers form the two main grades of inspecting staff. The inspectors (and regional and departmental directors) will usually be graduates. Controllers do not require a degree (although many have them). Graduate controllers can apply to join the inspector grade. The great majority of inspectors do not have a technical background but have legal, economic or labour relations training. About 20% of inspectors and 50% of controllers are women. There is no differentiation between inspectors who deal with general employment matters and health and safety issues.

22   The training of inspectors, both initial and continuing, is carried out at the Institut National de Travail (INT), a purpose-built training centre near Lyon. INT is currently administered by an inspector and the teaching staff are drawn from within the Inspectorate and from outside sources, both academic and industrial.

23   Labour inspectors have access to advice from the mines engineers (see paragraph 27), from the insurance companies (like CRAM – see paragraph 32) and various research and information organisations such as the INRS (see paragraph 44) and OPPBTP (see paragraph 50). They can also

require technical advice to be obtained by the employer.

24 Inspectors have rights of entry and can demand access to papers and information. They can impose obligations on employers to take certain health and safety measures and institute proceedings by presenting papers to the public prosecutor. Between 3000 and 4000 prosecutions are successfully taken each year. The decision whether to prosecute or not is the responsibility of the public prosecutor, based on the information provided by the inspector.

25 At local level inspectors are the sole permanent contact within a fixed geographical area for employers and employees and their representatives. They are autonomous and make their own decisions in terms of visits and priorities. In particular, it is the inspectors who draw up reports which are passed to the state prosecutor or who make decisions in the case of disagreement between the employer and employees' representatives. Regional directors only have a light co-ordination role. However, particular priorities may be set by the Ministry – in 1990 the priority for inspection was construction. In general, the inspectors aim to visit large employers once a year.

26 An annual report is produced by the Ministry of Labour which includes information on the number of visits and prosecutions made.

Mines engineers

27 Mines engineers are technically qualified specialists who, in addition to being responsible for mines and quarries, also cover environmental matters. They inspect 'classified premises' which include sites covered by Great Britain's Control of Industrial Major Accident Hazards Regulations 1984 and Notification of Installations Handling Hazardous Substances Regulations 1982. They also extend to lower hazard industrial undertakings which might have environmental risk implications.

28 The borderline between the labour inspector and the mines engineer is legally prescribed. Generally the labour inspector has no interest in what happens outside the factory and is concerned only for the safety of workers inside the factory. The mines engineer on the other hand is responsible for discharges from the factory, offsite safety and also for the integrity of the plant on site. The division of responsibility arose in 1977 as a result of legislation. The mines engineers that cover 'classified premises' work for the Environment Department but are paid by the Industry Ministry.

National Sickness Insurance Fund (CNAM)

29 The Caisse Nationale de l'Assurance Maladie (CNAM) – National Sickness Insurance Fund, is a bipartite organisation answerable to the Minister of Health and Social Security. It is financed by premiums levied on employers. It was established in 1947 (under the Social Security Code) to replace existing private arrangements for sickness insurance. Before then the private insurance systems offered a variety of levels of protection and the objective of the legislation was to develop equity in the level of protection. The legislation also required all sectors of industry to be covered by insurance. When the regime was originally established, employers had the choice to enter the national scheme or keep their existing arrangements but there is no scope for employers to opt out now, having made the original decision. Those employers who have exercised their right to have their own insurance include the state, some local authorities and large employers like the state bank and railway.

30 CNAM is the largest national insurance scheme but there are other smaller schemes covering other sectors of industry, for example, agriculture and professional/self-employed craftsmen. As well as controlling the operation of the sickness insurance funds the functions of CNAM are:

(a) to promote and encourage measures to prevent accidents and occupationally-induced ill health;

(b) to comment on laws and regulations proposed on occupational health and safety matters;

(c) to manage the National Fund for the Prevention of Occupational Accidents and Disease;

(d) to seek the best possible co-ordination with other parts of the safety system in France and its European equivalents;

(e) to participate in the development of standards which incorporate safety considerations;

(f) to administer the budgets of INRS (paragraph 44) and the preventive services of CRAM;

(g) to collate statistics relating to accidents and ill health at work.

It produces an annual report providing statistics and information on costs, activities and staffing levels.

31   The organisation is controlled by the National Advisory Committee with representatives of employers, employees, government and independent experts. This Committee is in turn assisted by fifteen Technical Committees spread across different sectors of industry. Each Technical Committee is composed of nine employer representatives, nine worker representatives and a similar number of their 'deputies'. They organise studies on risks at work and means of preventing them. These studies provide the basis for advice to the National Advisory Committee on matters such as prevention, statistics and insurance premiums. The committees also collate and analyse the accident and ill health statistics in their industrial sector and give advice to their equivalent regional committees. They also endorse the 'dispositions générales' (see paragraph 36) of Regional Technical Committees and extend them, in some circumstances, across the rest of the country.

Regional Sickness Insurance Funds (CRAMs)

32   There are 16 Regional Sickness Insurance Funds (Caisses Régionales d'Assurance Maladie – CRAMs) which are also bipartite (although since 1986 there have been more employee than employer representatives) and are controlled and financed by CNAM. Representatives of employees are elected by all employees covered by the insurance system in the region. The Regional Committees develop and co-ordinate measures to prevent ill health and accidents at work and apply the rules for the setting of premiums.

33   The activities of the CRAMs are carried out within the framework of the general policy drawn up by CNAM. This general policy can be modified to reflect any regional characteristics or guidance recommended by the Regional Technical Committees.

34   The CRAMs have a specialist technical service consisting of advisory engineers (200) and safety controllers (400). Both groups generally have an industrial background and technical experience. They also have laboratory technicians who can visit to take samples. These are based in various laboratories and centres throughout the country.

35   The advisory engineers and safety controllers of the CRAMs have the right to enter all workplaces which are subject to the general social security regime. They can carry out any measurements, analyses or atmospheric samples they consider necessary. They also investigate accidents – not to determine responsibility but to identify the cause in order to educate and advise on remedial measures.

36   CRAMs can use 'dispositions générales' which require all employers carrying out the same activity to take certain preventive measures. These dispositions must be submitted to the appropriate Regional Technical Committee and sanctioned by the Regional Director of Labour.

37   The Regional Committee can apply to the National Committee to seek approval to extend the 'dispositions générales' to the rest of the country.

38   CRAM engineers perform a preventive but essentially advisory role and do not make reference to legal texts to force an employer to take necessary action. They do however require employers to take all justifiable preventive measures. Where employers are unwilling to take the necessary action or the risks are of a serious nature they can increase premiums.

39   In the event of an accident, the employee will be compensated by CRAM for loss of earnings and any disability. Where there is evidence of severe negligence the employer can be required to reimburse the social security administration.

40   Employers' premiums are determined in a complex way. There is a policy to attempt to make the payments to CRAM reflect the performance of the employer. This is easier for large employers with large number of employees for whom it is possible to work out accident statistics and frequencies. However for smaller employers this is not practical. Consequently, for employers with less than 20 employees there is a general tariff per employee. For employers with more than 300 employees the premium is based on their accident record and performance over the last three years. There is then a sliding scale for employers with between 20 and 300 employees where the ratio between general and individual premium changes. There is however a bottom line and companies with no accidents or ill health still have to pay a basic premium.

41   Visits by advisory engineers and safety controllers are similar to those carried out by labour inspectors. They speak to employers and employees as well as representatives of the safety

committee. They prioritise their planned visits against a form of risk assessment depending upon the number of accidents per year and risks to health.

Regional Technical Committees

42   These committees are also bipartite and assist the CNAM/CRAMs. Each region has a different number of committees depending on the indigenous industry and the number of employees. The committees are consulted on proposals for increases or reductions in premiums by CRAMs. They also decide on any financial penalty applied to employers who have not carried out the recommendations of CRAMs advisory engineers. They can put up premiums by multiples of 25% up to a maximum of 200%.

43   One percent of the budget of the Regional Technical Committees is in the form of grants, made available to employers who have exceeded their obligations. Any such grants are improved by the Technical Committees.

National Safety and Research Institute (INRS)

44   The Institut National de Recherche et de Sécurité (INRS) – National Safety and Research Institute – was set up by legislation in 1947. It is funded by CNAM and therefore indirectly by compulsory contributions levied on employers. It is managed by a bipartite council with representatives of employers and employees. It has five main functions:

(a) to increase the awareness of health and safety issues at work;

(b) to carry out studies and research into occupational accidents and ill health;

(c) to gather and distribute documentation on health and safety;

(d) to co-ordinate the syllabus and methods of training of occupational health and safety 'professionals', safety engineers, technicians, occupational doctors, architects and engineers; and

(e) to provide a technical service for the Ministries of Labour and Social Security, the Sickness Insurance Funds, Technical Committees and medical inspectors.

45   Research represents 40% of INRS's work and is undertaken at their research centre in Nancy where they employ 400 people including 200

scientists. This centre covers a wide range of subjects including vibration, acoustics, robotics, ergonomics, industrial psychology, protective equipment, toxicology and epidemiology. The Labour Inspectorate and sickness fund engineers can send samples to the INRS laboratory for analysis.

46   INRS has a team of six occupational health doctors who maintain close contact with their peers in industry to keep them up-to-date on toxicological matters, and also to use them to monitor developments in industry.

47   The documentation service is comprehensive and INRS actively disseminates information to interested parties. They are changing their approach to dissemination, feeling that in the past it has been misdirected at big companies or specialists and not at small firms. (The larger companies represent 5% of the total number of companies in the country and employ 55% of the workforce, whereas 95% of companies are small and employ the other 45% of the workforce.) Targeting at small firms is being achieved by a series of initiatives – including an exhibition bus – as well as providing information in simplified packages. No charge is made for any of the publications produced in-house.

48   The library service employs 30 staff and has established a publicly accessible database which firms can access through the *Minitel* system. Only 500 of the 1.4 million companies use this system emphasising the need to get to small firms.

49   Clients can use the technical services information department and training department of INRS. The department also provides an information service by phone, letter or visit which answers something in the order of 20 000 enquiries per year.

Organisation for the Prevention of Accidents in Construction and Public Works (OPPBTP)

50   The Organisme Professionnel de Prévention du Bâtiment et des Travaux Publics (OPPBTP) – Organisation for the Prevention of Accidents in Construction and Public Works – was set up in 1947 by the Minister of Labour. It followed the introduction of legislation which required workplaces to have health and safety committees. Committees were required at that time for companies with more than 50 employees. It was considered inappropriate to have these committees for some industries (like construction and public works) where the mobility of staff and the transient nature of the workplace made the

arrangements difficult. OPPBTP was intended to fill the gap left in the absence of such committees in these industries. (It should be noted that public works includes civil engineering, construction and sectors of industry that are concerned with the manufacture of goods for the building industry and their subsequent use on construction sites.)

51   OPPBTP is a bipartite organisation, managed by a National Committee with equal representation from employers and employees. Its main function is to research the underlying causes of risks and hazards, propose measures to alleviate them, and inform and provide training for the professions involved. OPPBTP has no power over companies. It relies on persuasion and places great emphasis on its professionalism and the experience of its staff in the industry.

52   The organisation is financed by a levy payable by employers in the industry and calculated on the number of employees.

53   OPPBTP also has 16 Regional Committees (rather like CNAM/CRAM). Each Committee has representatives of employers and trade unions and includes a president and vice-president, a representative of the Ministry of Labour, a representative of CNAM, a general secretary and a medical adviser. The regions employ engineers, safety advisers, information specialists and administration staff. The headquarters is staffed by generalists, educationalists and trainers, policy staff and administrators. The engineers and safety advisers employed by OPPBTP must have a minimum of five years' experience in the building and public works industry and are provided with training in accident and ill health prevention.

54   OPPBTP employs about 365 staff, 75 of whom are engineers. Sixty staff work at headquarters and service the National Committee, 44 are at the Regional Training Centre and the rest are located in the regions.

55   The organisation provides a vast array of educational and audio-visual training aids, including videos. It publishes two journals, one for the profession itself and another for students, engineers and trainers. A small charge is made for documentation.

56   OPPBTP has a training centre at Saint-Jean-de-Braye near Orleans – the Centre Pierre Caloni (named after the founder) – which provides a wide programme of training courses in the construction and public works area. It was set up in 1972 and provides short courses on health and safety for the construction industry.

57   For the last 20 years OPPBTP has provided the secretariat of the construction section of the International Social Security Association (ISSA), set up some 60 years ago, whose headquarters is in Geneva. This groups social security associations from countries all over the world and provides technical co-operation and training courses in developing countries.

58   OPPBTP staff visit workplaces analysing the risks and advising on necessary remedial action. This normally has the desired effect but if not, they may refer employers to their trade associations which can use their influence to encourage change. Ninety percent of construction companies are member of trade organisations and they wield considerable power over their members.

59   If OPPBTP staff come across an imminent risk, they ensure the necessary action is carried out immediately. If necessary they call on the Labour Inspectorate. They also carry out technical investigations after accidents. The main purpose of this is not to apportion blame but to find out the underlying cause to propose remedial action. All accidents are required to be notified to OPPBTP.

60   For any construction project that costs more than 12 million French francs and involves more than ten sub-contractors, there is a requirement to set up a committee to co-ordinate planning of safety on the site. The committee will involve employers, employees and sometimes a delegate of OPPBTP. Also, since 1977, a site operational plan must be prepared and sent to OPPBTP. This enables OPPBTP to intervene and influence methods of work before construction begins.

61   OPPBTP also intervenes with suppliers and manufacturers. For example, in the development of proximity detection devices for tower cranes working in 'confined' areas and to get chemical manufacturers to replace dangerous chemicals in their products. The example of action on tower cranes came about as a result of a number of accidents caused by cranes colliding or toppling over. The problem was discussed by OPPBTP and INRS along with ANACT (see paragraph 64), manufacturers, users of cranes and employee representatives. The resulting solution – a stopping device – was agreed to by all and immediately implemented by the manufacturers. Much of this work is carried out in conjunction with ANACT and INRS.

62   OPPBTP has no public protection role.

63   OPPBTP has prepared safety guidance specifically aimed at small firms – there are some

200 000 construction employers with less than ten employees. The guides produced tend to be not about the law but about safety precautions (although trying to ensure that this reflects what the law says). OPPBTP also tries to influence safety in small enterprises by tapping into the highly organised apprentice system. There are approximately 40 000 apprentices per year going through various institutions and OPPBTP organises competitions on safety issues especially designed for apprentices.

National Agency for the Improvement of Working Conditions (ANACT)

64   The Agence Nationale pour l'Amélioration des Conditions de Travail (ANACT) – National Agency for the Improvement of Working Conditions – is a tripartite public body set up under the authority of (and funded by) the Minister of Labour. It was created in 1973 in response to concerns expressed during a period of industrial unrest over work conditions. It is governed by an administrative council consisting of representatives of employers and employees as well as the government and independent experts.

65   ANACT has three main functions:

(a) to gather and disseminate information on measures to improve working conditions. The information is distributed via various publications as well as forming the basis of an enquiry service. Conferences and meetings are organised to try to improve conditions of work and ANACT participates in international congresses;

(b) to provide a technical service to business – visiting and giving advice on health and safety implications of projects or plans. Such visits are at the request of industry – ANACT has no right of entry to premises. In providing advice, ANACT is keen to get the views of employers and employees. This makes for more effective advice and also ensures that ANACT is seen to be independent of management;

(c) to advise the Ministry of Labour on their Fonds pour l'Amélioration des Conditions de Travail (FACT) – Fund for Improvement of Working Conditions. Businesses make applications to the Ministry of Labour for financial assistance for a project that is designed to improve conditions at work or the environment. Bids are send to ANACT for evaluation who consider the projects on merit

and on the degree of involvement of the social partners.

66   ANACT has some 20 people employed on the information side and a number of other specialists including ergonomists, psychologists and engineers. Although split into different departments the organisation is flexible and able to intervene across disciplines. ANACT has five regional offices throughout the country with about four specialists in each. The head office at Montrouge, Paris has around 50 staff giving a total of 100 in the organisation as a whole.

67   ANACT has developed a 'diagnostic tool' (in effect a health and safety check list) which is made available to small companies to enable them to identify health and safety problems. The use of the diagnostic tool is free to businesses – it is paid for by the Labour Ministry.

68   If ANACT is invited to advise or help in a project, a member of staff will visit the company and spend two or three days speaking to management and staff and observing what goes on. They then produce a report for the company, identifying any problems and suggesting solutions. This consists of a series of broad brush proposals rather than an action list and may recommend that consultants be brought in to develop specific recommendations for action. ANACT has access to a network of consultants, built up over ten years, who are recognised for this type of work. The initial visit by ANACT is paid for by the Ministry of Labour but if consultants are subsequently used, the company is required to pay for them. However, they can apply to ANACT for financial support.

69   Although labour inspectors may refer people to ANACT it does not normally provide advice to inspectors – there is no direct working link between them. This is considered to be important because ANACT relies on the invitation and good will of employers and aims to encourage them to change their practices by persuasion and sound argument. The labour inspector on the other hand is seen as a law enforcer.

The Industrial Relations Directorate of the Ministry of Labour

70   In addition to the Labour Inspectorate there is another subdirectorate of the Ministry of Labour which is concerned with occupational health and safety. This is the Subdirectorate for Conditions of Work and Protection Against Occupational Hazards – Sous-direction des Conditions de Travail

et de la Protection contre les Risques du Travail which:

(a) develops legislation to protect workers from occupational hazards;

(b) oversees the enforcement of legislation;

(c) prepares rules concerning the organisation and operation of occupational health services;

(d) liaises with government departments and other organisations on matters concerned with the improvement of working conditions;

(e) participates in research;

(f) provides information and advice within the Ministry on questions concerning occupational health, pathology and ergonomics; and

(g) supervises and provides technical guidance for regional medical inspectors.

Higher Council for Prevention of Occupational Hazards (CSPRP)

71 The Conseil Supérieur de la Prévention des Risques Professionnels (CSPRP) – Higher Council for Prevention of Occupational Hazards – is a national council chaired by the Minister of Labour. It is an advisory and consultative body for the Ministry of Labour and includes representatives of the administration, employers, employees and other experts. It meets in plenary session once a year. It is supported by a smaller permanent commission (also chaired by the Minister of Labour), which meets once a quarter, and a number of specialist subject commissions, which meet more frequently. The specialist commissions cover:

(a) information, training and prevention;

(b) biological, chemical and physical environmental risks;

(c) physical, mechanical and electrical risks;

(d) occupational ill health;

(e) industrial medicine.

72 CSPRP also provides a national forum that enables co-ordination of the various players involved in health and safety at work to be considered at a policy and formal level

The Social Security Directorate of the Ministry of Health and Social Security

73 The Social Security Directorate of the Ministry of Health and Social Security is responsible for supervising the activities of the National Sickness Fund, the Regional Sickness Insurance Funds and the National Safety and Research Institute. It is also responsible for the prevention of occupational disease.

French Standardisation Organisation (AFNOL)

74 Standards development is the responsibility of l'Association Française de Normalisation (AFNOL). This organisation can commission other parties to produce standards on a contract basis or will set up its own committees to fulfil a remit.

Bibliography

1 *Factory inspection – historical development and present organization in certain countries* International Labour Organisation, Geneva, 1923

2 *The law and practice concerning occupational health in the member states of the European Community* Graham & Trotman Ltd, 1985

3 *La Prévention des Risques Professionnels – Les Acteurs – 1989* French Ministry of Labour

4 *La Direction des Relations du Travail* French Ministry of Labour, 1989

5 *Prevention of occupational hazards in France – structures and operation (Prévention des Risques Professionnels en France – Structures et fonctionnement)* Pietruszynski M, rev-ed, Paris, Institut National de Recherche et de Sécurité, 1986

6 *Handbook of labour inspection (health and safety) in the European Community* D E Clubley, Health and Safety Executive internal report, 1990

7 *Labour inspection, purposes and practice* International Labour Office, Geneva, 1973

8 *Comparative analysis of the reports of tripartite missions assessing the effectiveness of labour inspection systems in seven countries of Western Europe* International Labour Office, Geneva, 1985, ISBN 9 22 105242 7

9 *Labour inspection in the European Community* Stewart Campbell, Health and Safety Executive, HMSO, 1986, ISBN 0 11 883871 7

NATIONAL OCCUPATIONAL ACCIDENT STATISTICS: FRANCE

by Graham Stevens,
HSE's Statistical Services Unit

Introduction

1   This paper describes the systems of compensations for and reporting of injuries in France. It also presents some injury statistics for 1986 and 1987 and says how these compare with the injury record for Great Britain.

2   The French industry and severity classifications of injuries are governed by the determination of employer contributions to the insurance funds. Published injury figures suggest that Great Britain has substantially lower injury rates than France. This paper suggests that the difference is overstated, though with Great Britain still being lower.

Background to reporting system

3   The statistics on occupational injuries and diseases flow from compensated claims made by employees. Salaried employees are covered in occupational health insurance by a number of assurance funds, though the main fund – the National Sickness Insurance Fund – covers around 78% of employees.[1][2] This national fund, Caisse Nationale de l'Assurance Maladie (CNAM), is a national public body under the control of the Ministry of Social Services and has, since an act of 1946, been responsible for compensation of employees for workplace injuries and diseases. An organisation diagram is given in Figure 1.

4   There are separate insurance funds for workers in agriculture, railways, public bodies such as energy, and central and local government.

Insurance

5   The National Sickness Insurance Fund (CNAM), through its fund for salaried employees (CNAMTS), is responsible for furthering research into accident prevention. CNAM funds the National Institute of Research and Safety (INRS), which undertakes research into occupational hazards and provides technical assistance to labour inspectors. CNAM and INRS do not supplement the enforcement role of the labour inspectors who are employed by the Ministry of Labour.[1][3]

6   CNAM works with National Technical Committees (CTNs) for each of the 15 main branches of industry covered by the fund.[4] The CTNs, comprising of representatives from employers, employees and some from CNAM, submit strategies for prevention to the Ministry of Labour.

7   The national organisation of CNAM is partitioned at the regional level into a number of Caisses Régionales d'Assurance Maladie (CRAMs), and then into Local (Primary) Funds (CPAMs). The CRAMs keep registers of accidents, days lost due to injury, and certain details of employers in their region. The detail on employers is extensive – numbers of accidents, total payments made to victims, numbers of employees, salary payments, relevant CTN, and code of the national states register of establishments. By law companies must register with a CRAM.

8   The regional versions of the CTNs – CTRs, work with the CRAMs on accident prevention and can take legal action against employers, though not on enforcement but on paying additional premiums for unsafe working practices.

9   The CPAMs administer the claims and pay any resulting compensation. They forward their details to the appropriate CRAM. The CPAMs are really akin to local benefit offices and report to a regional head office. It is the CPAM that may contest a claim for compensation.

10   Employers alone are responsible for funding CNAM – not employees or the Government. The insurance premiums are assessed by Ministry of Social Services officials, with representations by the CTNs and their regional offices. In simple terms the premiums are broadly proportional to numbers of accidents and costs of compensation.

11   The premiums for 'large' (over 300 employees) employers are based on costs from individual premises; for 'small' (below 20) employers are based on industry group; and for 'other employers' on a sliding scale of both factors. The calculations can be complex[5] and involve past payments, numbers of employees and total extent of incapacity. A company/employer is assigned by the Regional Fund (CRAM) to an historical risk classification of industry groups, on the basis of activity of the majority of employees.[1] The kind or type of injury (save incapacity level) does not contribute to the derivation of premiums.

Definition and compensation of occupational injuries

12   The definition[1][2] of an occupational injury is an injury which arises out of or in connection with work where the employee is absent from work for at least one day after the day of the accident and is compensated by CNAM. A more literal translation taken from the French Labour Code is given in the appendix.

Figure 1 Organisation of system of accident reporting and compensation

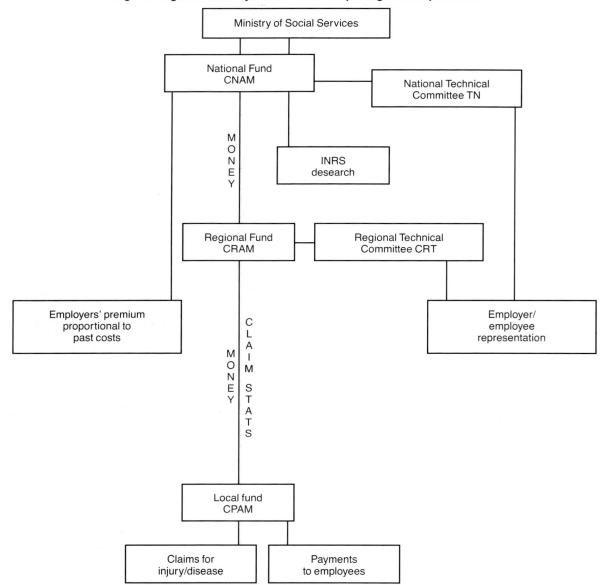

13   Injuries resulting in death before the fixing of the compensation or invalidity pensions are counted as fatal injuries. The pension is usually fixed within two to three months after the accident causing the injury.[4] Most deaths from work injuries occur before the pension is fixed. For example, in 1987 there were 1004 deaths from work injuries which occured before the pension was fixed, but only another 62 injuries resulted in death after the pension was fixed.[2] CNAM treat the 1004 deaths as the official work accident total for 1987.

14   Injuries are divided into two categories of incapacity: permanent and temporary (PI and TI). An injury with PI is one that leads to permanent disability or death. There is no closed definition of an injury with PI. However, an injured employee with PI will have had an amputation, fracture or injury which results in some loss of function [6] and which primarily determines the level of compensation. Ten percent of occupational injuries involve PI. In Great Britain 'major injuries' account for 11% of all-reported injuries but there is substantial under-reporting of over-3-day injuries. If only a third to a half of over-3-day injuries are reported in Great Britain then major injuries would account for only between 4 and 5½% of all reportable injuries. This would suggest that the French definition of a PI is wider than the definition of a major injury in Great Britain.

15   Commuting accidents are separately recorded and analysed by CNAM.[2] But an injury sustained by an employee on the road while travelling is treated as an occupational injury. This is because the employee was working when the injury occurred. These are called 'vehicles' – accidents that involve moving vehicles but not forklift trucks and large earth moving machines. Around 4% of work injuries are of this road type.

16   The compensation paid to injured employees is largely determined by the level of incapacity, with no compensation for under 24-hour accidents. For example in 1986 the average cost of an accident with PI was 133 856FF (about £15 000), while the figure for temporary incapacity accidents was 6390FF (around £700). There is no specific differential between payments for injuries and for diseases. Also no distinction is made in payment between commuting and work injuries.[7] Benefits to injured employees include – daily cash payments, medical treatment and any fixed sum awarded for an annuity.

Reporting of occupational injuries

17   The employer must complete an injury declaration (report) and send it to CPAM within 48 hours of knowing about an injury. The declaration is signed by the employer (manager) and the injured employee. The employee's own general practitioner also completes a declaration for CPAM, following examination of the employee.

18   CPAM can contest the work component of this claim up to 21 days following receipt of the declarations. Many claims [1][4] are rejected as not being occupationally linked. Heart attack cases are also rejected, usually being assessed by the emergency services and the employer's doctor.

19   On false claims, the demands on CNAM far outstrip the monies for compensation. As a result, the employer and employee declaration must give real proof of occupational link before payment is made.[4]

20   There is no guarantee that all non-fatal injuries are being reported.[1] For example in construction more work is increasingly being contracted out to self-employed workers – whose injuries are not counted in the statistics. Also employers, in order to reduce their insurance premiums, may be tempted not to report accidents. This, and the fact that claims undergo scrutiny, would suggest that not many non-work accidents get compensated, and hence counted in the work accident statistics.

Some injury statistics for 1987

21   CNAM publishes statistics on employees insured with the Fund and the compensated (over-1-day injuries). CNAM also publishes figures for other employees covered by some other special insurance regimes. For 1987 CNAM[2] published the injury and employment figures contained in Table 1 below.

22   The average number of days lost per work stoppage accident has risen steadily from just under 29 days in 1978 to over 33 days in 1987.

Comparisons between France and Great Britain

23   Incidence rates for France and Great Britain are given in Table 2. Some vehicle accidents are included in those from the special regimes (non-CNAM); and are around 7% of the fatal and ½% of over-1-day injuries. Numbers of injuries and rates for 1987 are given in Table 3. All figures exclude

TABLE 1 1987 Employee and injury statistics for CNAM and special regimes

| Insurance regime | Employees | Number of accidents with: | |
		work stoppage (a)	deaths
CNAM (all branches of industry)	13 305 883	662 800	1 004
CTN head offices	191 660	1 687	1
Electricity de France	142 569	2 942	9
Gaz de France	17 163	636	3
Mines	59 005	5 521	13
State Railway (SNCF)	219 480	9 140	30
Paris Transport (RATP)	39 977	2 085	
Building Society (Credit Foncier de France)	3 897	14	
Mutual agricultural associations	930 903	43 581	72
Totals	14 910 987	728 406	1 132

Note:
(a) Accidents excluding commuting industries. Of injuries reported to CNAM, the numbers of vehicles accidents are 22 391 over-1-day, and 424 fatal injuries. This leaves 706 015 over-1-day and 708 fatal injuries

commuting accidents. Further detailed breakdowns are given in the appendix. The comparison between France and Great Britain, in injury rates, is affected by differences in definitions of a reportable work accident and by reporting systems. As described below, these factors do not affect the comparison of fatal injuries.

24   The definitions of a work accident in France and Great Britain are similar in that the statistics include only those arising out of or in connection with work. Furthermore, as mentioned earlier, the treatment in France and Great Britain of the time to wait for a death following an accident is not likely to substantially affect the fatal numbers in France and hence the comparison.

TABLE 2 Injury incidence rates 1986, 1987 France and Great Britain (rates per 100 000 employees)

Industry	Fatal injury rates			
	Great Britain		France	
	1986/87	1987/88	1986	1987
Agriculture	8.6	6.8		7.7
Energy	5.8	6.7	9.9	10.5
Manufacture	2.1	1.9	3.8	3.5
Construction	10.2	10.3	16.5	17.5
All services	0.6	0.7	2.8	2.9
– transport	2.7	3.8	7.4	8.8
– other	0.4	0.3	2.1	2.1
All industries	2.0	1.9	4.5	4.7

Industry	Rates of all reported injuries			
	Great Britain (over-3-day)		France (over-1-day)	
	1986/87	1987/88	1986	1987
Agriculture	477	608		4 682
Energy	4 108	3 479	5 404	4 352
Manufacture	1 209	1 180	5 745	5 380
Construction	1 995	1 948	12 517	12 264
All services	341	370	3 114	3 045
– transport	998	1 032	6 357	6 383
– other	251	280	2 682	2 619
All industries	815	807	4 996	4 735

Notes for France:
1   Employees and injuries from the National Insurance Sickness Fund (CNAM) and most special insurance organisations
2   Public sector employees and injuries are not included. Thus local and central government, education and public health are not included
3   Manufacturing includes food distribution. Transport includes ambulance transport
4   The bulk of injuries come from CNAM and exclude vehicles accidents – those where the victim was injured by a moving vehicle. These occur mainly on roads. Such injuries are included in those from the special regimes; and are estimated to increase the all-industry rates by less than 0.5% for over-1-day and 7% to 8% for fatal injuries

25   An important point in comparing non-fatal injuries is that France counts over-1-day absences for work while Great Britain counts over-3-day injuries. Such a definitional difference may not be substantial since the time for employees to be away from work averages around 33 days per non-fatal injury. Informal sources in HSE suggest that where British employees are off work due to an accident they are then away from work for well over three days.

26   An examination of the Great Britain and French Industrial Classifications, has suggested a comparison by industry on a broad basis. There are some differences of detail; for example, manufacturing in France includes some repair and hiring of heavy goods which feature under services for Great Britain. The French service sector excludes public sector employees. In the comparisons of Table 1 French manufacture includes a small component of food manufacture associated with the food distribution industry. French manufacturing includes much of the food distribution trade. Some examination has shown that manufacturing rates are likely to be around 2 to 4% higher without food distribution.

27   Some of the difference in rates between France and Great Britain is due to the different industrial mix of the two countries. If France had the same distribution of employment by industry (at least to main industry level) as Great Britain in 1987/88 then the French fatal rate of 4.7 would drop to 4.2. This rate would drop further to 3.9 if we slice 8% from the number of deaths to allow for vehicle accidents in the special regime statistics. The rate of occupation fatal injury in Great Britain is then around half that in France.

28   For main industries, Great Britain has a lower fatal injury rate than France. This finding is unlikely to be affected by any differential under-reporting in the two countries. Earlier we noted that few non-work related accidents are counted in the French statistics. Also, under-reporting of British fatal injuries is unlikely because HSE gets to know of all work deaths from various sources, including coroners' reports. The overall fatal rate for France 1987 in Table 2 is 4.7, some 40% less than the published rate of 7.5.[2] These points suggest that the difference between fatal rates in the two countries is not as great as implied by the published figures, but that Great Britain is still lower.

Recent trends

29   The rate of fatal injuries has been declining over the past three decades although this long-term decline has been subject to occasional increases. The all-industry fatality rate, excluding vehicle accidents was 8.1 in 1978, 6.3 in 1983, 4.5 in 1986, 4.7 in 1987 and 5.1 in 1988.

TABLE 3 Employee occupational injuries and rates for main industry sectors excluding vehicle accidents

France 1987				
	Injury numbers		Rates	
Industry	Number of over-1-day injuries (a)	Number of deaths	Rates of over-1-day injuries	Rates of deaths (b)
Agriculture	43 581	72	4 682	7.7
Energy	11 969	29	4 352	10.5
Manufacture	279 284	183	5 380	3.5
Construction	148 936	212	12 264	17.5
All services	222 245	212	3 045	2.9
− transport	52 724	73	6 383	8.8
− other	169 521	139	2 619	2.1
All above (d)	706 015	708	4 735	4.7

GB 1987/88			
Industry	Number of deaths	Rate of over-3-day injuries (b)	Rates of deaths (b)
Agriculture	21	608	6.8
Energy	33	3 479	6.7
Manufacture	99	1 180	1.9
Construction	103	1 948	10.3
All services	77	370	0.7
− transport	48	1 032	3.8
− other	29	280	0.3
All above	342 (e)	807	1.9

Notes:
(a) Injuries for which the employee was off work for more than 24 hours, not including the day of accident, and where compensation was paid
(b) Rates expressed per 100 000 employees
(c) Excludes central and local government, and education
(d) Around ½% of over-1-day and 7% of fatal injuries involve moving vehicles. French manufacturing includes food distribution
(e) Includes 9 fatal injuries unclassified by industry.

30   The increase in the fatality rate between 1986 and 1988 also appears in the rate for transport services (up by 27%) and construction (up by 44%). There was also a 5% rise in the rate for agriculture between 1987 and 1988.

31   In 1988 there were also increase in the rate of serious (permanent incapacity) injuries in all main sectors except agriculture, punctuating a decline since 1983. In transport, the PI injury rate dropped from 591 per 100 000 employees in 1983 to 509 in 1987 and then rose 9% to 553 in 1988. The figures for construction are 1697 in 1988; 1318 in 1987; and 5% up, to 1388, in 1988. All these rates excluded vehicle accidents. Injury rates for all industries are given in Table 4, with British rates given for comparison; and serious injury rates for construction and transport in Tables 5 and 6. These show that the recent reversal in rates of serious injury in France was only mirrored in Great Britain in the construction and transport sectors and then by not as much. The fatality rate for Great Britain in 1988 was affected by the Piper Alpha disaster. This details the comparison.

32   All-reported injuries appear to have fallen over the period in both countries but the comparison may be affected by changes in the level of reporting of over-3-day injuries in Great Britain. With the exception of non-transport services, there appears to be a close correspondence between changes in serious injury rates between 1986 and 1988 and changes in employment; with sectors where employment rose seeing an increase in serious injury rates, while in sectors where employment was stable or falling the serious injury rate was either level or falling. The same relationship can be observed for Great Britain but appears to be weaker. In the construction and transport sectors, numbers of employees in France rose by only 3% and 2%, respectively, but in Great Britain they rose by 5% and 4%, respectively. Yet the serious injury rate rose by the same or a larger proportion in France than Great Britain in these sectors. In the manufacturing sector the serious injury rate was level in both countries over the same period although employment was up by 1% in Great Britain while it fell 2% in France.

33   The implication of this is that Great Britain appears to have been more successful than France in containing the adverse effects that economic expansion put upon health and safety resources and commitments of firms in the latter part of the 1980s, although both countries appear to have suffered from such effects to some extent.

TABLE 4 Fatal and serious injury rates in France and Great Britain – all industries except the public sector

	Fatal		Serious		All reported	
	Great Britain	France	(incl fatal) Great Britain	France	Great Britain	France
1983	2.5	6.3		606	N.A.	5 977
1986	2.0	4.5	100	467	815	4 996
1987	1.9	4.7	97	449	807	4 735
1988	2.8	5.1	96	466	794	4 743
Change	%	%	%	%	%	%
1983-88	+12	−19		−23		−21
1986-88	+40	+13	− 4	No change	−3	−5
1987-88	+47	+ 9	− 1	+4	−2	No change

Notes:
All rates exclude vehicle accidents and are expressed per 100 000 employees
The rates for 1986 onwards for Great Britain are for financial years beginning 1 April
The definition of a serious (major) injury in Great Britain was widened on the introduction of new reporting regulations on 1 April 1986. Serious injuries since then are not comparable with those previously reported. The new regulations also re-introduced the requirement to report other non-serious over-3-day injuries whose reporting fell into obeyance for a few years

TABLE 5 Fatal and serious injury rates in construction

	Fatal		Serious	
	Great Britain	France	Great Britain	France
1983	11.6	23.0		1 697
1986	10.2	16.5	293	1 355
1987	10.3	17.5	287	1 318
1988	9.9	23.8	296	1 388
Change	%	%	%	%
1983-88	−15	+ 3		−18
1986-88	− 3	+44	+1	+2

Notes:
All Rates are per 100 000 employees and excluding vehicle accidents occurring on the road. Rates for Great Britain for the years 1986 onwards are for financial years beginning 1 April

TABLE 6 Serious injury rates in transport in France and Great Britain (excluding vehicle accidents)

	Great Britain	France
1983		591
1986	92.1	511
1987	96.3	509
1988	99.1	553
Change	%	%
1983-88		−6
1986-88	+8	+8

Notes:
Rates for Great Britain from 1986 onwards are for financial years beginning 1 April

Appendix 1   Detailed breakdowns of employee numbers and injuries

TABLE A1 France year 1978
Employees and injuries of the National Sickness Fund (CNAM) and special regimes

Industry (a)	Number employees	Injuries		Rates	
		over-1-day (b)	Fatal	over-1-day (c)	Fatal
Agriculture	894 678	62 281	187	6 961	20.9
Energy	269 320	24 447	41	9 077	15.2
Manufacture	6 151 252	516 206	329	8 392	5.3
Construction	1 634 676	245 183	325	14 999	22.9
All services:	6 170 745	232 368	290	3 766	4.7
– transport	810 246	60 215	123	7 432	15.2
– other	5 360 499	172 153	167	3 212	3.1
All (d)	15 120 671	1 080 485	1 222	7 146	8.1

Notes:
(a) CNAM 15 main industry branches and special regimes. Energy includes mines. Manufacture includes food distribution. Services excludes public administration, education and health. Transport includes some ambulance services
(b) Injuries resulting in work stoppage of at least 24 hours (not including the day of injury) and where compensation was paid
(c) Rates per 100 000 employees
(d) CNAM injuries exclude vehicles/road traffic accidents to employees while working. A few traffic accidents are included in the special regime accidents. Their estimated percentage of the totals are 7% to 8% of fatals and less than ½% of all accidents

TABLE A2 France year 1983
Employees and injuries of the National Sickness Fund (CNAM) and special regimes

Industry (a)	Number employees	Injuries		Rates	
		over-1-day (b)	Fatal	over-1-day (c)	Fatal
Agriculture	959 774	53 928	100	5 619	10.4
Energy	291 638	24 760	23	8 490	7.9
Manufacture	5 772 028	414 298	279	7 178	4.8
Construction	1 407 200	196 449	323	13 960	23.0
All services:	6 876 984	225 462	234	3 279	3.4
– transport	845 592	57 405	84	6 789	9.9
– other	6 031 392	168 057	150	2 786	2.5
All (d)	15 307 624	914 897	959	5 977	6.3

Notes:
(a) CNAM 15 main industry branches and special regimes. Energy includes mines. Manufacture includes food distribution. Services excludes public administration, education and health. Transport includes some ambulance services
(b) Injuries resulting in work stoppage of at least 24 hours (not including the day of injury) and where compensation was paid
(c) Rates per 100 000 employees
(d) CNAM injuries exclude vehicles/road traffic accidents to employees while working. A few traffic accidents are included in the special regime accidents. Their estimated percentage of the totals are 7% to 8% of fatals and less than ½% of all accidents

TABLE A3 France year 1986
Employees and injuries of the National Sickness Fund (CNAM) and special regimes

Industry (a)	Number employees	Injuries		Rates	
		over-1-day (b)	Fatal	over-1-day (c)	Fatal
Agriculture					
Energy	281 576	15 216	28	5 404	9.9
Manufacture	5 281 782	303 447	203	5 745	3.8
Construction	1 229 789	153 937	203	12 517	16.5
All services:	7 081 135	220 516	195	3 114	2.8
− transport	832 996	52 954	62	6 357	7.4
− other	6 248 139	167 562	133	2 682	2.1
All (d)	13 874 282	693 116	629	4 996	4.5

Notes:
(a) CNAM 15 main industry branches and special regimes. Energy includes mines. Manufacture includes food distribution. Services excludes public administration, education and health. Transport includes some ambulance services
(b) Injuries resulting in work stoppage of at least 24 hours (not including the day of injury) and where compensation was paid
(c) Rates per 100 000 employees
(d) CNAM injuries exclude vehicles/road traffic accidents to employees while working. A few traffic accidents are included in the special regime accidents. Their estimated percentage of the totals are 7% to 8% of fatals and less than ½% of all accidents

TABLE A4 France year 1987
Employees and injuries of the National Sickness Fund (CNAM) and special regimes

Industry (a)	Number employees	Injuries		Rates	
		over-1-day (b)	Fatal	over-1-day (c)	Fatal
Agriculture	930 903	43 581	72	4 682	7.7
Energy	275 033	11 969	29	4 352	10.5
Manufacture	5 191 392	279 284	183	5 380	3.5
Construction	1 214 392	148 936	212	12 264	17.5
All services:	7 299 454	222 245	212	3 045	2.9
− transport	825 991	52 724	73	6 383	8.8
− other	6 473 463	169 521	139	2 619	2.1
All (d)	14 910 987	706 015	708	4 735	4.7

Notes:
(a) CNAM 15 main industry branches and special regimes. Energy includes mines. Manufacture includes food distribution. Services excludes public administration, education and health. Transport includes some ambulance services
(b) Injuries resulting in work stoppage of at least 24 hours (not including the day of injury) and where compensation was paid
(c) Rates per 100 000 employees
(d) CNAM injuries exclude vehicles/road traffic accidents to employees while working. A few traffic accidents are included in the special regime accidents. Their estimated percentage of the totals are 7% to 8% of fatals and less than ½% of all accidents

TABLE B1 France CNAM employees and injuries (injuries omitting vehicles [EM08]) year 1986

Main industry activity	CNAM group number	Employees	Accidents	
			over-1-day	Fatal
Energy (water, gas, electricity) (a)	(13)	51 867	2 902	2
Manufacturing (b)	(1, 3-11)	5 281 782	303 447	203
Construction	(2)	1 229 789	153 937	203
All services: – transport (c) – other (d)	(12) (14, 15)	6 613 795 562 817 6 050 978	207 607 41 425 166 182	169 42 127
Total		13 177 233	667 893	577
Vehicle accidents		—	22 709	401
Total for CNAM		13 177 233	690 602	978

Notes:
(a) Includes quarrying
(b) Includes food distribution
(c) Includes ambulance transport
(d) Mainly business/offices and non-food distribution

Other insurance regimes

	Employees	Accidents	
		over-1-day	Fatal
Agriculture			
Public transport Paris and SNCF	270 179	11 529	20
Public gas/electricity/mines	229 709	12 314	26
Offices of CINS	197 161	1 380	6
Grand total (CNAM and other)	13 874 282	715 825	1 030

TABLE B2 France CNAM employees and injuries (injuries omitting vehicles [EM0H]) year 1987

Main industry activity	CNAM group number	Employees	Accidents	
			over-1-day	Fatal
Energy (water, gas, electricity) (a)	(13)	55 846	2 870	4
Manufacturing (b)	(1, 3-11)	5 191 205	279 284	183
Construction	(2)	1 214 392	148 936	212
All services: – transport (c) – other (d)	(12) (14, 15)	6 844 440 566 534 6 277 906	209 319 41 499 167 820	181 43 138
Total		13 305 883	640 409	580
Vehicle accidents			22 391	424
Total for CNAM		13 305 883	662 800	1 004

Notes:
(a) Includes quarrying
(b) Includes food distribution
(c) Includes ambulance transport
(d) Mainly business/offices and non-food distribution

Other insurance regimes

	Employees	Accidents	
		over-1-day	Fatal
Agriculture	930 903	43 581	72
Public transport Paris and SNCF	259 457	11 225	30
Public gas/electricity/mines	219 187	9 099	25
Offices	195 557	1 701	1
Grand total (CNAM and other)	14 910 987	728 406	1 132

40

TABLE C Industrial Classification

List of French industries, as classified by French insurance, that are contained in tables (A, B) for comparison with Great Britain.

Industry	Insurance organisation
Agriculture	Mutual agricultural associations
Energy	CNAM Branch 13 Gaz de France Electricitè de France Mines
Manufacture	CNAM Branches 01, 03-11
Construction	CNAM Branch 02
Services Transport	CNAM Branch 12, SNCF, RATP (State Railway, Paris Transport)
Other	CNAM Branches 14, 15 Crèdit Foncier de France (Building Society) CTN Head Offices

Main Branches of Activity

Source: CNAM

01 Metallurgy
02 Building and public works
03 Timber
04 Chemical
05 Non-metallic minerals
06 Rubber, paper, cardboard
07 Books
08 Textiles
09 Clothing
10 Leather and skins
11 Food (+ food distribution)
12 Transport and handling
13 Water, gas, electricity (excluding EDF-GDF)
14 Commerce (wholesale and retail trade – non food)
15 Interprofessional (some wholesale and business and research)

Definition of a work injury translated by Mons Bastide of the INRS

"An occupational accident is an accident which occurs because of the work done or duing work, whatever the cause, and involves any salaried person or any person working in any way and at any place for one or seveal employers or firm managers" (Labour Code, art L 411-1)

Bibliography

1   National Statistics of Accidents at Work (France): in *Systems for monitoring working conditions associated with Health and Safety (Systèmes de suivi des conditions de travail liées à la santé et à la sécurité)* Telle M A, Dublin, European Foundation for the Improvement of Living and Working Conditions, 1990

2   *National Statistics of Work Accidents 1985-86-87.* Also 1988 publication for the years 1984-85-86 Caisse Nationale de l'Assurance, Maladie des Travailleurs/Salariés

3   *Organisation chart of Accident Prevention in France* Institut National de Recherche et de Sécurité, 1982

4   Telle M A (private communication April 1990)

5   *Tarification des Risques d'Accidents de Travail et de Maladies Professionelles* Reverchon J, Institut National de Recherche et de Sécurité 1989

6   Christofari M F, Ministry of Labour: Statistics (Private communication April 1990)

7   Bastide J C, Institut National de Recherche et de Sécurité (private communication May 1990)

REPORT ON THE SYSTEM FOR THE HEALTH AND SAFETY PROTECTION OF WORKERS IN WEST GERMANY

By Sally Van Noorden and Julia Soave,
Health and Safety Executive

Introduction

1   This report was written by officials of the Health and Safety Executive in June 1990. At the time of the report, the unification of West and East Germany was the top item on the political agenda and plans for economic union from July 1990 were well advanced. The timetable for political union was less clear, but it was evident that the expectation was that the five East German federal states would join the 11 West German federal states in due course. Plans were in hand for extending the West German worker health and safety system to East Germany, and discussions between officials were taking place about harmonising accident insurance arrangements and other social security measures.

2   Consequently this report concentrates on describing the West German system at the time of writing and does not venture into speculation about future developments.

Legal and institutional framework

General features of the system

3   West Germany's law concerning health and safety at work derives from two sources: public labour law and social insurance legislation. The most general principle of their labour law is contained in the Industrial Code (Gewerbeordnung), dating from 1891, which creates a general obligation on employers to 'arrange and maintain workplaces, plant, machinery and tools and to run the business in such a way that the workers are protected against dangers to life and health so far as the nature of the business permits'. The Code empowers the Federal Government to issue ordinances or orders (Verordnungen) containing more detailed provisions. Further laws based on the legal framework of the Industrial Code, and a number of other acts on worker protection, have also been passed. The federal states may also promulgate their own laws and orders.

4   The social insurance legislation, which dates from 1884, is contained in the State Insurance Code (Reichsversicherungsordnung). This provides for workers' accident insurance funded by employers and administered by accident insurance associations (Berufsgenossenschaften) which are managed jointly by employers' and employees' representatives. The Code requires the accident insurance associations 'to use all appropriate means to prevent accidents at work' and in particular to issue regulations. These accident prevention regulations

(Unfallverhütungsvorschriften) are ratified by the Federal Ministry or the federal states (Länder) and are legally binding on the members of the appropriate insurance association. In general, they differ from federal legislation in that they deal with specific machines and technical processes rather than social questions and major hazards affecting a wide spectrum of employment.

5   West Germany has a dual system for the enforcement of these two groups of legislative provisions. First, there is a Labour Inspectorate (Gewerbeaufsichtamt) organised on regional lines by the Labour Ministries of the 11 federal states. Second, there is a Technical Inspectorate (Technische Aufsichtsbeamte) employed by the accident insurance associations to enforce their regulations and advise member enterprises on safety. Both Inspectorates date from the late nineteenth century.

6   The Federal Ministry for Labour and Social Affairs (Bundesministerium für Arbeit und Sozialordnung) has responsibility for developing national policy and legislation on labour protection, and handles co-operation between the two Inspectorates by administrative instruction.

7   The Federal Ministry is advised by an institute called the Federal Institute for Occupational Safety and Health (Bundesanstalt für Arbeitsschutz – BAU)* which advises the Ministry on technical matters; produces guidance material for labour protection practitioners; runs seminars and conferences and develops learning packages for use in colleges; conducts its own research and also commissions research from outside bodies; and maintains an occupational safety and health information centre.

8   The industrial accident insurance associations have their own active research and testing centre known as the BG Institute for Occupational Safety (Berufsgenossenschaftliches Institut für Arbeitssicherheit – BIA). They run well-established and extensive programmes of safety training within firms; and take the lead in making testing arrangements for equipment safety standards.

9   There are also private institutions within the health and safety system. Of particular importance are 11 Technical Inspection Associations (Technische Uberwachungs – Vereine) which are independent technical

*The abbreviation BAU is used because the Institute was formerly called the Bundesanstalt für Arbeitsschutz und Unfallforschung.

consultancies and which are recognised under the Industrial Code to perform certain statutory inspections of hazardous plant.

10   Equally important is the German Standards Institute (Deutsches Institut für Normung – DIN), a private corporate body which draws up standards for the design of plant, machinery and equipment, and also for their safe use. Its governing body includes representatives of government, employers, trade unions and consumers. The technical standard-setting work is carried out by a number of representative specialist committees and is largely in response to requests from clients who may be government, industry, or the accident insurance associations. There are also other standard setting bodies, such as the Association of German Electrotechnicians (Verband Deutscher Elektrotechniker – VDE).

11   At the level of the firm, certain health and safety arrangements are statutorily required. An undertaking normally has a Works Council (Betriebsrat), which is a body of elected worker representatives with statutory rights to consultation and co-determination on a wide range of industrial relations issues. Firms also have a Labour Protection Committee (Arbeitsschutzausschuss), which is a committee of all those concerned with health and safety and which is required by law to meet at least once a quarter.

12   Employers are required to appoint safety specialists (Sicherheitsfachkräfte); works doctors (Betriebsärzte); and safety stewards (Sicherheitsbeauftragte). Safety specialists are employees within a firm at engineer, technician or foreman level whose role is to advise management on safety questions. Employers are required by law to appoint and train a certain number of full or part-time specialists depending on the size and nature of the enterprise. The precise requirements are determined by the accident insurance associations and set out in their regulations. Works doctors are appointed to advise employers and employees on occupational health and the provision of first aid; only very large companies are required to have a full-time works doctor. Safety stewards are appointed by the employer from amongst rank and file employees; their role is to check compliance with safety regulations.

General principles of the law

13   As was explained in paragraph 3, West Germany's law concerning health and safety at work derives from two sources: public labour law and social insurance legislation. These two groups of legislative provisions will be described separately.

14   The general principle contained in the Industrial Code (see paragraph 3) has been supplemented by orders containing more detailed provisions, in particular to protect employees and third parties against the hazards of certain dangerous installations (eg boilers, other pressure vessels, lifts) by requiring these to be regularly inspected and licensed. Such installations are known as 'installations requiring inspection' (Uberwachungsbedürftige Anlagen). Orders based on the Industrial Code include the Work Places Order (Arbeitsstättenverordnung) and twelve on 'installations requiring inspection'.

15   Since the Industrial Code, a number of other acts on labour protection have been passed. These go considerably wider than health and safety at work and include legislation on hours of work and protection of mothers and young people. There is separate legislation concerning mining and nuclear safety and the safety of explosives. The main legislation concerning hazardous substances (which is the responsibility both of the Ministry of the Environment and of the Labour Ministry) is contained in the Chemicals Act 1980 (Chemikaliengesetz), supplemented by hazardous substances orders. Other important health and safety at work acts are the Works Doctors, Safety Engineers and other Safety Specialists Act 1973 (Arbeitssicherheitsgesetz); and the Safety of Equipment Act 1968 (Gerätesicherheitsgesetz), under which the GS (geprüfte Sicherheit) symbol (the equivalent of the kitemark) is established.

16   Some acts and orders empower the Government to issue a further layer of provisions containing detailed guidance on implementation. These provisions may take the form either of general administrative regulations (Allgemeine Verwaltungsvorschriften), or of guidelines (Richtlinien).

17   Acts, orders, general administrative regulations and guidelines also contain references to 'technical rules' (technische Regeln). These are standards, *inter alia*, for the manufacturing design and safe use of plant, machinery, materials, substances, equipment and the working environment. The purpose of referring to such standards is to enable the law to keep pace with developments in science and technology. Some federal laws refer to specific standards, but it is more usual to require the employer or manufacturer to take account of relevant standards in general. The most commonly used term is 'generally acknowledged rules of

Thru plc
Photos pages from
R.A Document of
colour photocopier in
Camden High St.

technology' (allgemein anerkannte Regeln der Technik). For example, the Safety of Equipment Act stipulates that manufacturers and importers may only put technical equipment on the market if it has been produced in accordance with the generally acknowledged rules of technology. These are understood to be technical rules which are generally acknowledged by experts in the field; kept up-to-date with scientific and technological advances; and drawn up by a procedure in which all experts and other interested parties have an opportunity to participate.

18   Another category of standards referred to in federal legislation is 'established principles of labour science' (gesicherte arbeitswissenschaftliche Erkenntnisse). The Work Places Order and the Hazardous Substances Order require employers to take account of both generally acknowledged rules of technology and established principles of labour science. The latter include subjects such as work psychology and ergonomics. Other terms used in federal law for standards include 'the state of technology' (Stand der Technik) and 'the state of science and technology' (Stand von Wissenschaft und Technik).

19   Acts containing a general requirement to observe standards sometimes help the employer to identify which standards meet that requirement. For example, the Work Places Order states that the relevant standards are in particular to be taken from the Work Places Order guidelines. Standards pertaining to the Safety of Equipment Act are listed in appendices to the Act's general administrative regulations. Standards pertaining to the Hazardous Substances Order are issued by a special multi-representative committee formed by the Federal Labour Ministry.

20   The accident prevention regulations of the industrial accident insurance associations are, as mentioned in paragraph 4, concerned with specific hazards. These regulations are drafted by specialist committees (Fachausschüsse) set up by the Central Association of Industrial Accident Insurance Associations (Hauptverband der gewerblichen Berufsgenossenschaften). Once draft regulations have been agreed, they are submitted to the Federal Ministry, which is required to consult the federal states before giving approval. Regulations are promulgated by the individual accident insurance associations. They are supplemented by enforcement instructions, directed at the insurance associations' technical inspectors (see paragraph 5).

Extent of coverage of the dual system

21   Most health and safety at work provisions under public labour law apply only to commercial and industrial enterprises (Gewerbe). Excluded from coverage are the whole of the public sector (including the police, post office, railways, army, publicly owned hospitals and educational institutions), agriculture and forestry, seafaring and charities.

22   However a few provisions contained in federal labour protection law do apply to all enterprises in the economy, in particular the 'installations requiring inspection'; the Chemicals Act and Hazardous Substances Order; and the legislation on the protection of mothers and young people.

23   With regard to social insurance law, most employees are entitled to insurance against accidents at work and associated accident prevention measures. There are accident insurance associations similar to those in industry for agriculture, mining and seafaring; and equivalent accident insurance bodies for the three tiers of the public sector (municipalities, the Länder and federal institutions). However, at the time of writing (June 1990), certain groups of employees were not statutorily covered by accident insurance and associated inspection arrangements, although some were in fact covered by analogous voluntary and advisory arrangements. These groups were civil servants; church employees; and employees of NATO forces stationed in West Germany. As a result of the requirements of the European Community Framework Directive, arrangements will have to be made by the end of 1992 for these groups of people to be statutorily entitled to health and safety at work protection; and the Federal Ministry is giving consideration as to how this should be effected.

The dual inspection arrangements

24   As was explained in paragraph 5, West Germany has a dual system of health and safety at work inspection. First, there is a state Labour Inspectorate; and second, there are technical inspectors employed by the accident insurance associations.

25   The state Labour Inspectorate is not a federal institution but is organised by the Labour Ministries of the 11 federal states. However, there are a number of bodies which co-ordinate the activities of the Labour Inspectorate and provide for an exchange of information, both at the political and the professional level.

26   In most federal states the Labour Inspectorate is responsible for environmental protection (Unweltschutz) as well as labour protection, and it is estimated that the environmental work takes up to 50% of the Inspectorate's time. The labour protection work covers (as indicated in paragraph 15) not only work on health and safety but also work on hours and protection of mothers and young people. The health and safety work takes up only 10 to 15% of the Labour Inspectorate's time.

27   The Labour Inspectorate is responsible for enforcing all state law on labour protection. It has no direct power to enforce the accident prevention regulations issued by the accident insurance associations, but can use them as a source of information for enforcing the general duties on employers laid down in the Industrial Code. Inspectors can issue an enforcement notice (Anordnung), which orders the employer to take specific measures within a specified period to remove or reduce a hazard. They are also able, under the Safety of Equipment Act, to prohibit manufacturers or importers from selling or displaying a piece of equipment.

28   If an employer fails to comply with an enforcement notice, the Labour Inspectorate can impose an administrative fine (Bussgeld). In cases of very serious accidents, the Inspectorate can refer the case to the public prosecutor (Staatsanwaltschaft) for action under criminal law.

29   The technical inspectors of the industrial insurance associations are, in the main, accident specialists in particular industries. They have no responsibility for environmental protection and no right to enforce state law. Their duties are to enforce the accident prevention regulations of the insurance associations and to advise their members generally on accident prevention.

30   The technical inspectors have similar powers of enforcement to the state labour inspectors, but these are founded in social law and cases would ultimately go to the social court rather than the administrative court. Like the state inspectors the technical inspectors can ultimately refer a case to the public prosecutor if they believe a criminal offence has been committed.

Occupational health services

31   Both the federal states and the accident insurance associations provide occupational health services which have qualified specialist doctors. These doctors are in addition to the works

doctors required at the level of the firm (paragraph 12).

The duties of inspectors

The Labour Inspectorate

32   In most federal states the Labour Inspectorate operates from a number of offices, organised on either regional or sectoral lines. There are about 3400 inspectors altogether, who fall into three civil service grades: 'upper', for which a university degree is required; 'senior', for which a polytechnic qualification is needed; and 'middle', for which a technician or foreman certificate is needed.

33   Training of inspectors is the responsibility of the federal states. In most states, it lasts two years, and is pitched at appropriate levels for the three grades. Advanced and refresher training is available.

34   As explained (in paragraph 26) health and safety matters take up only 10 to 15% of the Inspectorate's time. Inspectors have rights equivalent to those of the police to enter and inspect places of work without prior notice; but in practice tend to give notice of visits. The policy on frequency of visits varies between federal states. Generally, the criteria used for determining the frequency of visits are the degree of danger, accident rate and size of firm; but decisions are made in the knowledge that the technical inspectors are likely to be giving the most hazardous undertakings a great deal of attention.

35   The Inspectorate does not make great use of legal sanctions; only 900 administrative fines and 50 prosecutions are recorded for 1988*.

The technical inspectors of the accident insurance associations

36   The accident insurance associations are run jointly by employers and trade unions. The financial contributions to these bodies are paid by the employers.

37   There are 95 accident insurance associations which fall into three groups: 35 industrial associations; 19 agricultural associations; and 41 public sector associations.

*This information comes from the annual report from the Federal Ministry of Labour and Social Affairs on Labour Protection for 1988 (entitled *Arbeitssicherheit 89*).

38   There are some 2000 technical inspectors. They have roughly the same conditions of service as the state labour inspectors and are graded in the same way.

39   The technical inspectors specialise in particular industrial sectors and they have other duties besides routine inspection, for example accident investigations and all external training of those concerned with labour protection at the level of the firm.

40   The accident insurance associations have an important sanction at their disposal connected with their insurance role. The State Insurance Code provides that the contribution which an employer has to pay to the appropriate accident insurance association may be varied according to the accident level in the firm, and therefore according to the demands made on the insurance association's compensation funds. The levels of premiums therefore vary considerably and each accident insurance association has its own arrangements for calculating appropriate surcharges and rebates.

The arrangements for occupational health services

41   As explained in paragraph 31, both the federal states and the accident insurance associations provide occupational health services.

42   There are 92 state medical inspectors. They are teams of qualified occupational health physicians organised by the federal states to provide an advisory and inspection service in occupational medicine. Their organisation is distinct from that of the labour inspectors.

43   The industrial accident insurance associations organise the provision of medical treatment and rehabilitation for those injured at work; and to this end manage their own specialist clinics staffed by their own doctors.

44   Employers have a general duty, laid down in the Industrial Code, to safeguard the health of their employees. This has been supplemented by specific regulations such as the state's Hazardous Substances Order and the accident insurance associations' Regulations on Carcinogens. Both the state legislation and the insurance associations' regulations require regular medical examinations for a number of occupational groups, eg those who work with certain hazardous substances. The examinations must be conducted by a doctor authorised by the state medical inspector in conjunction with the relevant accident insurance association. The authorised doctor is usually the works doctor, except in small firms.

45   There are a number of prescribed occupational diseases, and the accident insurance associations must do all they can to protect workers from these diseases.

46   As explained in paragraph 12, employers are required to appoint works doctors except in small firms. The required qualifications and numbers of works doctors are laid down in accident prevention regulations of the insurance associations. These provide that more and better qualified doctors are required in larger and more hazardous undertakings. Small firms are able to use the services of group occupational health services run by the accident associations, the employers' associations or the technical inspection associations.

Bibliography

1   *Report of the tripartite mission on the effectiveness of labour inspection in the Federal Republic of Germany* International Labour Office, Geneva, 1984

2   *The law and practice concerning occupational health in the member states of the European Community* prepared by Environmental Resources Ltd for the Commission of the European Communities, vol 2, Denmark and the Federal Republic of Germany, Graham and Trotman, 1985, ISBN 0 86 010627 6

3   *Labour inspection in the European Community* Stewart Campbell, Health and Safety Executive, HMSO 1986, ISBN 0 11 883871 7

4   West German contribution to *Handbook of labour inspection in the European Community* advance copy of text sent by the Federal Ministry to the European Commission, January 1990

5   Annual report from the West German Federal Ministry for Labour and Social Affairs covering worker protection in the year 1988, published in 1989

6   Information booklets describing their work from the German industrial accident insurance associations and from the Federal Institute for Occupational Safety and Health, June 1990

NATIONAL OCCUPATIONAL ACCIDENT STATISTICS: WEST GERMANY

by Graham Stevens,
HSE's Statistical Services Unit

Introduction

1   This paper describes how injuries are reported in the Western half of the Federal Republic of Germany. It also compares West Germany with Great Britain in respect of work injury record for some recent years. Standard published injury figures suggest that Great Britain has substantially lower rates of work injury than West Germany. The comparisons of this paper suggest that the two countries are much closer together in fatal injury rates, though with Great Britain still being lower. Over the latter part of the 1980s the gap between the two countries' rates of fatal and serious injury appears to have narrowed.

2   Currently published statistics on occupational injuries show rates of injury to be much lower in Great Britain than in the Western half of the Federal Republic of Germany.[4][9][10] This paper describes the system of reporting injuries in West Germany and the associated role of the accident insurance associations. It also presents some injury statistics for West Germany 1988 and says how they compare with the injury record for Great Britain.

3   Occupational injuries attract compensation and free medical treatment in West Germany. The link to compensation and medical treatment suggests that under-reporting of occupational injuries is negligible in West Germany. Also the assessment on claims for compensation and the variable insurance premiums from employers preclude large scale over-reporting. The fatal injury figures for 1987 and 1988 suggest that, after allowing for road fatalities, Great Britain has a lower rate of fatal injury than West Germany for some industries and for industry overall. The difference in the overall fatal rate is less than that implied by standard official figures from the International Labour Organisation (ILO)[10] and West Germany,[4][9] but is still present when some allowance is made for hours of work and part-time working.

Insurance

4   The statistics on occupational injuries and cases of ill health flow from claims for compensation made for, or on behalf of, insured workers. Employees and some self-employed are insured by the accident insurance associations (Berufsgenossenschaften-BGs) which were set up by the Reich Insurance Code (RVO) of 1911. The main point is that work is an insured activity so that individual workers must insure themselves

and any dependents for accidents or ill health that are not work related.

5   The main duties of the insurance associations for work related injury or ill health are:

   (a) provision of cash and pension benefits associated with compensation;

   (b) provision of medical treatment for injured and ill people;

   (c) rehabilitation; and

   (d) prevention of accidents at work (the associations make their own regulations, which have legal force, and have their own Technical Inspectorates)

6   The associations (BGs) have historically been structured into three main groups – industrial, agricultural, and independent. The industrial BGs cover manufacturing, construction, and the energy industries and business and non public services. Agriculture, forestry and fishing are covered by the agricultural BGs. Public services including government administration and education are covered by the independent BGs, but public gas, electricity and water are covered by the industrial BGs.

7   By law[1][2] employers must be registered with an insurance association, and pay premiums to cover compensation and other benefits associated with work injuries and ill health. A newly started business must register with a local authority who informs a central body, the federal state. This body determines which BG is appropriate for the new business. In 1988 over 1.97 million employers/companies were registered with the industrial BGs, 1.72 million with the agricultural BGs and nearly 365 000 with the independent BGs. The latter includes federal, state and city authorities, and also private households with employees.

8   Even employees of an unregistered company are covered by the insurance. In general, employees know of the insurance arrangements. In addition, subcontractors and their employees are insured. Thus 'members of the public' are not covered for injury due to work activity. In conclusion, work injuries are not likely to go unreported as a result of unregistered companies.

The reporting system

9   Once an injury has occurred at work, the injured goes to a doctor (not the works doctor) for

49

treatment and a certificate to say how long any absence from work might be. The injured person (or relative) must present this certificate to the employer to receive wages (for up to six weeks). The insurance association pays for the medical treatment including the doctor's fee, and also any wages after six weeks.

10   If the certificate specifies an absence longer than three days, then the injury must be reported to the association. Both employer and doctor complete a report for the association. Therefore a reportable work injury is broadly defined as an injury arising out of work:

(a) in the exercise of occupational activity (or connected with insured activity);[2][3]

(b) where the injured is absent from work for more than three days, not counting the day of accident or return.

11   The main points of this system are that injured employees must go to a doctor to obtain wages from the employer and that the doctor sends on the details of the injury with the bill to the insurance association. It is likely that non-recording of employee work accidents by the associations will be negligible.[2]

12   Commuting and other traffic injuries are reportable work injuries because people were going about their work. Not all the self-employed are required to register with an association.[1][2] A crude comparison for 1986 of those registered with the numbers from the European *Labour Force Survey* (LFS) suggests that nearly two thirds of the non-agricultural self-employed are registered. Some self-employed work injuries may not be recorded.

13   In addition to the employees and the self-employed, the BGs also insure workers who help or assist at public disasters (road accidents) or act in an honorary capacity (charity work).[1] Each year the BGs estimate how many such people act in this way – they are simply listed as 'other' (sonstige) in insurance reports.[1] The number of injuries suffered by the others is reckoned to be very small[2] but are not readily identifiable, the duration of the cover of insurance (exposure to this work activity) is reckoned to be low.[1][2]

14   Could some non-work accidents be compensated by an insurance association? There are two points in answer to this question. Firstly; the associations do assess the work-related nature of the claim based on the reports from the doctor

and employer. In some cases the association may need to go to court to fight a claim for a serious or fatal work injury. Serious and fatal injuries and cases of death through ill health related to work attract lifelong pensions (to the victims or dependants). These represent the largest item of expenditure by the insurance associations.[1] Thus the incentive is present for the associations to examine claims for compensation, particularly those arising from serious injuries and also from illnesses.

15   In the case of a fatality on the work premises, the employer must immediately contact the association and the police and medical emergency services. The medical services will establish the cause of death (natural causes or otherwise).

16   Secondly, the associations have (and use) the power to adjust the contributions from an employer according to the level of compensation from past accidents and/or ill health. There is no incentive for an employer to connive with a false claim for compensation due to a non-work injury.

Serious and fatal injuries

17   The reportable injury is the 'over-3-day' injury, and is called 'Meldepflichtige arbeitsunfalle'. Some of these are of the category of serious injury – first-time compensated (erstmals entschadigte arbeitsunfalle). In turn, some of these are fatal injuries (erstmals entschadigte todliche arbeitsunfalle).

18   The category of serious injury is where the victim has at least a 20% reduction in fitness for work and lasting longer than 13 weeks after the accident. The determination of the percentage reduction in fitness is complex.[1][2] Serious injuries attract permanent compensation or pension. Serious injuries are in no way comparable with the British definition of major injury. First-time compensated injuries account for around 2% to 3% of all-reported injuries in West Germany while major injuries account for 11% of all-reported injuries in Great Britain and, if only a third to a half of over-3-day injuries are reported, between 4 and 5½% of all reportable injuries. Thus the definition of a serious injury in West Germany appears narrower than the definition of a major injury in Great Britain.

19   A serious injury is counted as fatal if death occurs before the award of the pension.[1][3] There are no figures for the number of deaths occurring after the fixing of a pension. A special study by the Ministry of Labour[2] showed that over 50% of deaths occurred within 3 days of accidents, over

80% within 4 weeks and very few above 12 months. Since the time taken to fix a pension is at least 13 weeks after the accident the number of unrecorded deaths is thought to be very small.[2]

Some injury statistics for 1988

20   This section gives some published[1][4] employment and injury figures for the Federal Republic of Germany, 1988, together with some previously unpublished statistics[5][6] on injuries which exclude road traffic accidents.

21   Statistics on insured workers and their injuries are published by the Industrial Group of Associations and the Ministry of Labour and Social Affairs. The Ministry publications[1][4] give consolidated figures for all groups of insurance associations (BGs) – industrial, agricultural and independent. From these sources, in 1988, 24.87 million workers were insured with the industrial BGs, and of these, 23.63 million are dependent employees (abhangig beschaftigte). The bulk of the 4 million insured agricultural workers are self-employed or 'employers' (unternehmer). The independent insurers, for the public sector, covered nearly 3 million employees. Some further details are given in Table 1. The total in employment is nearly 32 million.

TABLE 1: Employment in F R Germany

Group of insurance association	Numbers of insured workers		
	Self-employed	Employed	Total
A Industrial	1 245 144	23 625 424	24 870 568
B Agriculture	3 014 667	976 500	3 991 167
A + B	4 259 811	24 601 924	28 861 735
C Independent (public sector)		2 993 741	2 993 741
All above	4 259 811	27 595 665	31 855 476

% Change in Employment between 1986/88 compared with Great Britain

	West Germany (all insured workers) %	Great Britain (employees in employment) %
Agriculture	− 3%	− 5%
Energy	− 5%	− 9%
Manufacturing	+ 9%	+ 1%
Construction	+12%	+5%
Transport	+ 1%	+ 4%
Other services	+ 6%	+ 7%

22   As explained in paragraph 13, the Ministry also publishes estimates of insured workers who help in voluntary activities. These are national

insured workers engaged in activities of short duration. They are thus excluded from the employment figures of Table 1, and from the denominators of the injury incidence rates of the next section. The industrial associations cannot readily distinguish the few injuries arising from the voluntary activities. Their effect on injury rates are negligible.[2] These few injuries are included in the injury figures of Table 2.

23   The BGs distinguish the 'scene' of an accident in the resulting six combinations of: accidents at the workplace, on the way to another workplace, commuting; and of road traffic, or non traffic.

24   Counted in non-traffic accidents are those involving fork-lift trucks and reversing lorries in warehouse loading bays. The nearest equivalent to British reportable injuries are the two combinations: non-traffic at work and on the way to another workplace. The latter categories are small in number. The numbers of road-traffic and non-road traffic work injuries and commuting injuries are displayed in Table 2.

25   Road traffic injuries are called 'strassenverkehrunfallen' and, together with commuting accidents 'wegunfalle', are removed from the West German injury figures for comparison with Great Britain. For the industrial and agricultural BGs, this means omitting 436 fatal injuries out of 1493. Thus 1057 non-traffic work fatalities are identified in Germany in 1988 for all industries excluding public administration and education. The figure for West Germany in 1987 is 1020.

Incidence and frequency rates

26   Injury incidence rates for West Germany are displayed in Table 3 for industry activities of the industrial and agricultural insurance associations. The incidence rates are based on injuries that exclude commuting and traffic injuries. The numbers of insured workers, employees and self-employed, form the denominators of the rates. The injuries are to both these groups of workers and include a few injuries to workers engaged in a voluntary or honorary capacity (paragraph 22).[5][6] Nevertheless, for 1988 the all-industry fatal rate of 3.7 is much lower than the official published rate[4] of 0.06 fatals per 1000 full-time workers (6 per 100 000). The reasons are mainly two fold – removal of workroad traffic fatalities and expression of rate per insured worker not full-time equivalent.

TABLE 2 Employment in F R Germany

Group of insurance associations	Numbers of reported injuries					
	All reported (a)			Fatal		
	non-road	road	Total	non-road	road	Total
Industrial	1 214 273	20 361	1 234 634	779	351	1 130
Agricultural	173 150	2 387	175 537	278	85	363
Total	1 387 423	22 748	1 410 171	1 057	436	1 493
Independent			168 824			112
All work injuries			1 578 995			1 605
Commuting in all associations			174 202			755
Total			1 753 197			2 360

Note:
(a) Injuries leading to more than 3 days' absence from work (not including the day of accident or return)

TABLE 3 Reported injuries in F R Germany

Industry	1988				1987	
	Injuries to insured workers (a)		Injury rates (b)		Fatal (c)	
	All reported (over-3-day)	Fatal (c)	All reported	Fatal (c)	Injuries (a)	Rate (b)
Agriculture	123 150	278	4 338	7.0	286	7.1
Energy (d)	17 945	77	5 515	23.7	43	12.9
Manufacturing (e)	637 707	278	5 479	2.4	271	2.5
Construction	226 170	221	9 139	8.9	253	10.2
All services (f):	332 451	203	3 187	1.9	167	1.7
– transport	51 233	84	5 644	9.3	71	8.1
– other	281 218	119	2 953	1.2	96	1.0
All above	1 387 423	1 057	4 807	3.66	1 020	3.69

Notes:
(a) Injuries to employees, the self employed and the few injuries to these groups who act in any voluntary capacity. Injuries exclude commuting and traffic accidents
(b) Injury rates per 100 000 employees and self-employed
(c) Fatal injuries before a pension or compensation has been granted
(d) Energy is mainly mining and extraction of metallic ores for F R Germany, together with production of gas and water. Oil extraction is included in manufacturing for FRG but in energy for GB. The fatal injury figure includes 51 deaths in the Stolzenbach pit disaster of June 1988. The rate, excluding these deaths would be 8.0 per 100 000 employees and self employed
(e) Includes hotel, catering and butchers' shops
(f) Services excluding public administration and public transport

27   Details of the industrial classification are given in Appendix 1. Manufacturing, construction, energy and non-public sector services are covered by the industrial insurance associations. The hotel/catering and butchering activities cannot be distinguished from food manufacture, and are counted under manufacturing, whereas they are counted under services in Great Britain. The rates for manufacturing would probably rise marginally if hotels/catering were excluded. On this basis, manufacturing accounts for over 40% of employment in West Germany.

28   Agriculture and construction cover broadly the same activities in the two countries. The service sector of West Germany in this paper excludes hotels/catering, the public sector (ie public administration, education, federal transport and postal services). The British service sector covered in this paper excludes public administration and education. However, these two service sectors are broadly comparable since the vast bulk of employment will be in common industries.

TABLE 4 Frequency rates of fatal injury in F R Germany

Industry (a)	1988			1987	
	Fatal Injuries (b)	Frequency rate (c)	Fatal injuries (b)	Frequency rate (c)	
Energy	77	0.162	43	0.087	
Manufacturing	278	0.016	271	0.016	
Construction	221	0.071	253	0.083	
All services:	203	0.011	167	0.010	
– transport	–84	0.057	71	0.051	
– other	–119	0.007	96	0.006	
All in industrial insurance associations	779	0.020	734	0.019	
Agriculture (d)	278	0.092	286	0.096	
	278	0.033	286	0.034	
Total	1 057	0.025	1 020	0.025	
	1 057	0.022	1 020	0.022	

Notes:
(a) Industrial classification as in Table 4 and in Table A1 of Appendix 1
(b) Fatal injuries excluding commuting and road traffic fatalities
(c) Rate of fatal injury per million hours worked
(d) Hours of work not directly published for agriculture. Two estimates are derived: one from coverting the official full-time worker total to hours worked; the other from the *International Labour Force Survey 1986 (LFS)*. For 1988 the full-time worker method gives a rate of 0.092 and the *LFS* method gives a much lower rate of 0.033

29   The available figures[1][4] preclude comparing the energy sectors of the two countries on the same basis (Appendix 1).

30   Frequency rates, expressed per million hours worked, are given in Table 4. Hours of work are taken from Ministry of Labour reports.[1][4] These reports publish hours of work for all industrial insurance associations and some public sector associations but not for agriculture workers. However, an estimate of agricultural hours can be made from the number of full-time workers also published in the reports. For details see Appendix 2 on hours of work. In addition, an estimate can be derived from the International *Labour Force Survey*[8] with an assumption about the number of working weeks in a year. The frequency rate associated with both methods of estimating hours of agricultural work are given in Table 4.

31   The resulting range of frequency rates places agriculture as a higher risk industry than manufacturing or services – the same finding occurs for incidence rates. This suggests that the true agriculture frequency rate is likely to lie between the limits given in Table 4.

Comparisons between West Germany and Great Britain

32   This section compares West Germany with Great Britain by injury incidence rates and by frequency rates. We then take into account the hours worked in both countries for main industry groups. Tables 5 and 6 display injury incidence rates for fatal and all-reported injuries; during 1988 in West Germany; and during 1987/88, 1988/89 in Great Britain. Table 5 also gives fatal injury rates in West Germany in 1987. The British figures exclude public administration and education. The West German figures exclude the public sector including transport and postal communication. Employment details are given in Table A2 of the Appendix 1 on industrial classification. Table 7 gives the corresponding frequency rates for fatal injuries.

33   The rate of fatal injury in non-public sector industries overall is less in Great Britain than in West Germany. The British 1988/89 rate, including Piper Alpha deaths, stood at 2.8, the 1987/88 rate was 1.9. The figure for West Germany in 1988 is 3.7 (though this is lower than the corresponding French value of 4.7 in 1987).

34   This comparison of incidence rates will not allow for hours of work or part-time working which is a crude measure of hours. Thus we must turn to the frequency rates of Table 7 for a comparison that allows for hours of work. The overall fatal frequency is less in Great Britain than in West Germany, though this relative advantage is around 6% less than that on incidence rates. The extent of part-time working as given by the *Labour Force*

TABLE 5 Fatal injury incidence rates for Great Britain and the Federal Republic of Germany

Industry	Fatal injury rate (a)			
	GB		FRG	
	1987/88	1988/89	1987	1988
Agriculture	6.8	7.0(e)	7.1	7.0
Energy (b)	6.7	42.7	12.9	23.7
Manufacture (c)	1.9	1.8	2.5	2.4
Construction	10.3	9.9	10.2	8.9
Services (d)	0.3	0.5	1.7	1.9
All above (d)	1.9	2.8	3.7	3.7

Notes:
(a) Rates per 100 000 workers, employees for GB, employees and self-employed for FRG
(b) Energy in FRG includes extraction of metallic ores, but does not in GB. Fatal injuries for 1988/89 GB includes those 167 deaths in the Piper Alpha disaster, and for FRG includes 51 deaths in the Stolzenbach pit disaster of June 1988
(c) Includes hotel, catering and butchers' shops in F R Germany
(d) Excludes all public administration, education and public transport in FRG. Excludes public administration and education in GB
(e) The rate of self-employed and employees combined was 8.5 in 1988/89

TABLE 6 Injury incidence rates for all reported injuries for Great Britain and the Federal Republic of Germany

Industry	Rate of injury (a)		
	GB		FRG
	1987/88	1988/89	1988
Agriculture	608	653	4 338
Energy (b)	3 479	3 198	5 515
Manufacture (c)	1 180	1 239	5 479
Construction	1 948	1 928	9 139
Services (d)	370	391	3 187
All above (d)	807	794	4 807

Notes:
(a) Rates per 100 000 workers, employees for GB, employees and self-employed for FRG
(b) Energy in FRG includes extraction of metallic ores, but does not in GB
(c) Includes hotel, catering and butchers' shops in F R Germany
(d) Excludes all public administration, education and public transport in FRG. Excludes public administration and education in GB

Survey (LFS)[8] suggest that the West German incidence rate should be reduced by 5% – counting a part-time employee as half a full-time employee.

35   So, allowing for hours of work, the fatal rate for all industries is genuinely less in Great Britain than West Germany. A further point on rates for industries combined is the industrial mix. Higher 'risk' industries constitute a larger proportion of the West German mix of industries than they do of Great Britains. Some employment details are given in Table A3 of Appendix 1 on industrial classification. If we apply the British 1988/89 industrial mix to the West German main industry fatal rates of Table 5 then the aggregate fatal rate become 3.08, dropping from 3.66. The

corresponding figures for 1987 are 2.81 dropping from 3.69. The British aggregate (1.9 in 1987/88) is still lower even if a further 5 to 6% is sliced from the West German fatal rate to allow for hours of work.

36   For manufacturing and services (not public sector) both incidence and frequency rates of fatal injury are less in Great Britain than in West Germany. Of the difference between the two countries in manufacturing (2.5 FRG – 1.9 GB), at most 10% is explained by differing compositions of manufacturing. Appendix 1 looks at this in some detail. The comparison for agriculture is complicated by the employed component differing in extent between the two countries and the limited comparison by frequency rate. West German agriculture, mainly worked by self-employees, has a fatal incidence rate little different

TABLE 7 Frequency rates of fatal injuries for Great Britain and the Federal Republic of Germany

Industry (b)	Frequency rate (a)			
	Great Britain (c)		F R Germany (d)	
	1987/88	1988/89	1987	1988
Agriculture	0.039	0.040	0.034 – 0.096	0.033 – 0.092
Energy	0.038	0.245	0.087	0.162
Manufacuture	0.011	0.010	0.016	0.016
Construction	0.059	0.056	0.083	0.071
Services	0.002	0.003	0.010	0.011
All above	0.012	0.018	0.022	0.022

Notes:
(a) Expressed per million hours worked
(b) Notes on industry as in Table 5
(c) GB hours of work are based on average weekly hours from the *1988 British Labour Force Survey* and an assumed lower limit of 41 working weeks per year. Thus frequency rates for GB represent an upper limit
(d) For FRG hours of work based on published sources. Agricultural hours have been estimated from conversion of full-time workers and also from the *1986 International Labour Force Survey*. The corresponding frequency rates for 1988 are 0.092 and 0.033

from the employee rate in Great Britain, but is less than the rate (8.5) for employees and self-employees combined. The frequency rate in agriculture lies within a range. The British frequency rate lies within this range. This suggests a difficult comparison where, at worst, agricultural rates are slightly higher in Great Britain than West Germany, but at best they are much the same.

37   In construction, incidence rates and frequency rates give opposite results –Great Britain is lower in frequency rate but higher in incidence rates. This reflects the fact that the European *Labour Force Survey* (LFS)[8] shows average weekly hours in construction are higher in Great Britain than in West Germany. More importantly, the yearly hours per worker in West German construction was 1250 in 1988 (derived from Tables A2 and A4 of the appendices). The *1988 Labour Force Survey* in Great Britain estimated that construction employees worked 42.8 usual hours per week. This translates to a yearly total of 1755 hours assuming a 41 week year, or to 1250 hours assuming a very short working year of 29 weeks. A tentative conclusion, is that British construction does have a lower rate of fatal injury than West Germany – when measured against hours of work.

38   The comparison of all-reported injuries – mostly over-3-day for Great Britain and West Germany – is heavily dependent on the known under-reporting of over-3-day injuries in Great Britain. The level of this under-reporting is not known precisely but has been crudely estimated at 50%.[9] For main industry sectors we see from Table 6 that the all-reported rates are substantially less in Great Britain than in West Germany – being one sixth for all-industries combined. We saw earlier that under-reporting in West Germany is likely to be minimal.

39   In the absence of any firm estimate of British under-reporting, the comparison of all-reported rates of injury will remain speculative. We can note, however, that the level of under-reporting would need to be over 80% for the rates of injury to be at the same level in both countries. This figure would drop to 65% if we make a further assumption that the ratio of non-fatal to fatal injury numbers is the same in both countries. Either way, for the true level of injury to be worse in Britain than in Germany, under-reporting in Britain would need to be very much higher than currently estimated.

Recent trends

40   Throughout the 1980s, rates of both fatal and serious (ie fatal and first-time compensated) injury have declined in all main industrial sectors. However a major pit disaster contributed to a near doubling of the fatal rate in the energy sector between 1987 and 1988. (In Great Britain this sector also saw the Piper Alpher disaster which occurred in the same year).

41   During 1986 to 1988, rates of serious injury (excluding vehicle accidents) fell by 11% overall but fell by 14% in construction (see Table 9). The rate for transport services rose marginally by 2% but examination of published rates (which include road vehicle accidents) per full-time worker suggest this may be a blip in a long-term decline (see Table 10).

42   The rate of all-reported (over-3-day) injuries continued to fall over this period – whether expressed per capita or per full-time equivalent worker – although in construction the decline seems almost to stop.

43   The decline in injury rates between 1986 and 1988 occurred at a time when employment was rising in all sectors except agriculture and energy (see Table 1). However, there are tentative signs that the rate of decline in injury rates may have slowed down.

44   The comparisons with Great Britain point to a somewhat better rate of safety improvement in West Germany over the past few years than in Great Britain. They show that fatal injury rates, excluding road vehicle accidents, fell by 37% between 1983 and 1987 in West Germany compared to a fall of 24% in Great Britain. For both countries, trends over the more recent period are complicated by disasters in the energy sector in 1988.

45   Trends in serious injury rates probably provide a truer reflection of underlying safety performance than trends in fatality rates or in all-reported injuries. The former being affected by single large events while the latter are difficult to interpret because of the scale of under-reporting of over-3-day injuries in Great Britain. Between 1986 and 1988 the all-industry rate of serious injuries fell by 11% in West Germany compared to a fall of 4% in Great Britain.

46   Within the construction and transport sectors a similar picture of impressive West German safety performance emerges. Serious injury rates in construction in West Germany

TABLE 8 Rates of fatal, serious and all reported injuries from 1983 to 1988 in West Germany and Great Britain

For all industries except public sector	Fatal		Serious		All reported injuries	
	West Germany	Great Britain	West Germany	Great Britain	West Germany	Great Britain
1983	5.9	2.5	191		5 700	
1986		2.0	156	100	5 375	815
1987	3.7	1.9	149	97	4 965	807
1988	3.7	2.8	139	96	4 807	794
Change	%	%	%	%	%	%
1983-87	−37	−24	−22		−13	
1986-88		+28	−11	−4	−11	−3
1983-88	−37	+12	−27		−16	

Notes:
Rates expressed per 100 000 employees in Great Britain and per 100 000 insured workers in West Germany, excluding road vehicle accidents (the rate for serious injuries in West Germany in 1986 is based on an estimate of the number of non-vehicle accidents)
Rates for 1986 onwards in Great Britain are for financial years beginning 1 April
All reported injuries are those leading to more than 3 days' absence from work in GB and Germany
Serious injuries in GB are those that lead to deaths or major injuries; in Germany they are injuries which are compensated for disability pension for the first time

TABLE 9 Fatal and serious injury rates in construction

	Fatal			Serious		
	West Germany (a)	(b)	Great Britain	West Germany (a)	(b)	Great Britain
1983	13.6		11.6	353		
1986		14.0	10.2	320	382	293
1987	10.2	15.6	10.3	266	365	287
1988	8.9	13.7	9.9	275	371	296
1989		13.9	9.4		351	308

Notes:
All Rates are per 100 000 insured workers in West Germany and per 100 000 employees in Great Britain
(a) Excluding road vehicle accidents, the figure for serious injuries in West Germany in 1986 is based on an estimate of numbers of non-vehicle accidents
(b) Including vehicle accidents. West German Rates are per 100 000 full-time equivalent insured workers

dropped (by 14% excluding vehicle accidents or 3% including vehicle accidents) between 1986 and 1988 and, although there was a small rise between 1987 and 1988, the 1989 published figure (which includes vehicle accidents) continues the long-term decline. In contrast, serious injury rates in construction in Great Britain have risen each year since 1987/88.

47 In the transport sector, West Germany's long term decline in serious injury rates was halted in 1988 by a small rise but the published figure for 1989 suggests that the downward trend has resumed. While the serious injury rate in transport rose in West Germany by 2% between 1986 and 1988, the serious injury rate in transport in Great Britain rose by 8% over the same period.

TABLE 10 Serious injury rates in transport

Year	Germany (a)	(b)	GB (a)
1983	213		
1986	182	231	92.1
1987	199	249	96.3
1988	186	235	99.1
1989		220	100.0
Change	%	%	%
1986-1988	− 2	2	8
1983-1988	−13		
1987-1988	− 7	− 6	3

Notes:
Rates are expressed per 100 000 employees for GB; per 100 000 insured workers in Germany
(a) Rates exclude vehicle accidents. The figure for Germany 1986 is based on the estimated number of non-vehicle accidents
(b) Rates exclude vehicle accidents. Rates for Germany are expressed per 100 000 full-time insured workers
State railways are included in GB transport but excluded in Germany

Conclusion

48 The preceding sections have considered the likely main factors comparing injury rates – fatal and non fatal – between Great Britain and West Germany. Under or over-reporting of fatal injuries are likely to have a small effect on the comparisons; and commuting and road traffic accidents have been eliminated from the comparison. In broad terms, the comparison is of the same sorts of work injury in both countries and in reasonably similar main industry sectors. Given this, the injury analysis with an allowance for hours of work, suggests that Great Britain does have a lower 'rate' of fatal injury than Germany – for industries overall and for some main industries. Examination of accident rate trends, particularly for serious injuries, over the latter part of the 1980s suggest that this gap is narrowing.

Appendix 1

German industrial classification

1 The industrial classification is derived from historical boundaries between the accident insurance associations (Berufsgenossenschaft BG) which were established by law in 1911.[1][4] There are three main groups of association – BGs – industrial; agricultural and independent. The industrial BGs, covering around 78% of insured workers, include activities such as mining, manufacturing, construction, transport and distribution. The public sector, including the federal railways and post are covered by the independent BGs. Public utilities, however, are covered by the industrial BGs.

2 At a very broad level, the BGs and their industrial activities can be aggregated to the main industry sectors of the GB Standard Industrial Classification (SIC). Inevitably there are some BG activities which cannot be set against the right main SIC sector. For example, hotels, catering and the meat trade are inseparable from food manufacturing and are thus counted in manufacturing for Germany. Such activities would be counted under services by the British SIC.

3 In addition, the energy sector in West Germany comprises mines, gas and water and extraction of metallic ores. Energy in the British SIC would not include metallic ores extraction but would include electricity production and sale, and oil extraction. Thus the energy sectors are not really comparable.

4 The full details are given in Table A1. Employment figures for main industry sectors are given in Table A2. The percentage distribution of

TABLE A1 Approximate correspondence between industries

GB SIC	F R Germany BG Nearest equivalent BGs	
Agriculture (0)	Federal mutual organisation of agriculture BGs	
Energy (1)	BG1	Mining – coal mining, extraction, potash and metallic ores
	BG4	Gas/water – production and supply of gas, water and electricity
Manufacture (2-4)	BG2	Quarry opencast extraction of non-minerals, oil
	BG3	Ceramics, production of
	BG5	Forge/rolling
	BG6	Engineering
	BG7	Iron/steel
	BG9	Precious metal
	BG10	Precision engineering electronics
	BG11	Chemicals
	BG12	Timber, paper products
	BG15	Printing
	BG16	Leather textiles
	BG18	Food manufacture (includes hotels/catering)
	BG19	Abattoirs and butchers' shops
	BG20	Sugar refining
Construction (5)	BG21 – 28 Construction in all federal states	
Services: – transport (7)	BG32	Transport and railway (not federal)
	BG33	Road vehicles
	BG34	Sea and inland shipping
– other (6,8,9)	BG29	Wholesale distribution and deposit stores/warehouses
	BG30	Shops/retail
	BG31	Administration – business & office services (not in the public sector)
	BG36	Private health service
	Independent BGs – public administration and education/health	

TABLE A2 Employment by appropriate SIC activities

F R Germany 1988				
Industry	Self-employed	Employed	Total	Total for 1987
Agriculture	3 104 667	976 500	3 991 167	4 031 983
Energy (a)	30	325 370	325 400	334 295
Manufacture (b)	442 787	11 195 793	11 638 580	10 733 072
Construction	149 930	2 324 719	2 474 649	2 484 394
All services:	652 397	9 779 542	10 431 939	10 090 663
– transport (c)	– 105 787	– 801 925	907 712	875 489
– other (non public sector) (d)	– 546 610	– 8 977 617	9 524 227	9 215 714
Total excluding public sector	4 259 811	24 601 924	28 861 735	27 674 317
Public sector	N.A.	2 993 741	2 993 741	2 971 431
All industries	4 259 811	27 595 665	31 855 476	30 645 748

Notes:
(a) Includes coal mining, extraction of metallic ores but not open cast extraction and public sector utilities
(b) Includes hotels/catering and meat retail trade
(c) Private bus and railway, road vehicles, sea and inland shipping. Excludes public transport and warehouses associated with road haulage
(d) Mainly business and office services, non-meat retail and wholesale distribution

TABLE A3 Distribution of Employment (%) by main industry sector: Great Britain and F R Germany

Industry	Distrubution by percentage		
	GB 1988/89 Employees	Germany 1988 Employees and Self-Employed	Germany 1988 Employees
Agriculture	1.6	13.8	4.0
Energy	2.6	1.1	1.3
Manufacture	28.2	40.3	45.5
Construction	5.6	8.6	9.4
Services	61.9	36.1	39.8
Total excluding public sector	18 192.7 (000s)	28 861.7 (000s)	24 601.9 (000s)

employment in the main industry sectors for Great Britain and West Germany are given in Table A3. West Germany employs more people in manufacturing and construction but less in services.

Composition within manufacturing

5   Within manufacturing, British and West German employment are distributed differently. Manufacturing, as in the GB industrial classes, can be partitioned into two main groups. These correspond to SIC's 2/3 and 4 and are: mineral extraction, metal manufacturing, chemicals and engineering and; food, drink, timber products, clothes and miscellaneous production. These two groups can be considered as 'heavy' and 'light' manufacturing. From the *1986 Labour Force Survey*, 69·2% of employment in West German manufacturing was in the heavy group. The figure

for the UK is 59·2%. These figures are based on consistent classifications of industry obtainable from the *Labour Force Survey*.

6   In order to assess what effect this has on the comparison for manufacturing we can apply the GB manufacturing mix to the West German rates. That is to standardise the West German fatal injury rate for GB manufacturing composition. One problem is that the West German industrial classification within manufacturing is not quite comparable with the GB SIC groups 'heavy' and 'light'. One way to overcome this is to take the *Labour Force Survey* employment figures (1986), for which comparability does exist, and apply West German *Labour Force Survey* employment to GB fatal rates. The result is that the GB fatal rate in manufacturing becomes 1.91 increasing from 1.85 (1.9 in Table 5 on page 54). This increase is only around 11% of the difference between GB and West Germany in manufacturing fatal rates.

7   Of course, it is still possible to standardise the West German manufacturing fatal rate for the GB manufacturing composition. This can be done by applying the West German mix that is nearest to the heavy and light categories of manufacturing. The heavy category corresponds to activities under insurance associations 2 and 3 to 11 of Table A1. The light category corresponds to insurance associations 12 to 20, as the light category does include hotels and catering activities. This will help explain why the percentage of West German manufacturing in the heavy industry, 62·9% in 1988, is less than the *Labour Force Survey* figure of 69·2% in 1986. In

1988/89, 58·9% of GB manufacturing was in the heavy activities.

8   The resulting standardisation to the GB mix lowers the West German manufacturing rate (1988) from 2.39 down to 2.36. This is a small change which, because of the hotel/catering point, is expected to be smaller than the adjustment to the British manufacturing rate seen earlier.

9   The conclusion is that the difference in fatal manufacturing rates between West Germany and Great Britain is not adequately explained by differences in sectoral composition.

Appendix 2

Hours of work and part-time working

Sources

1   Figures on hours of work related to insured employees and the self-employed are given in the reports of the Federal Ministry of Labour[1][4] and the Federal Association of Accident Insurance Associations.[7] These reports give hours of work for the industrial associations and for some public sector associations. No figures are given for agricultural associations.

2   The industrial associations cover the energy, manufacturing and construction industries not in the public sector. In these industries, a total of 38 655.77 million hours were worked by employees and the self-employed. This figure includes 1 760.38 million hours estimated to have been worked by these workers in a voluntary capacity.

Agriculture

3   Estimates of hours of work in agriculture can be found from two sources – published full-time worker numbers[1][4] and the international Labour Force Survey. Each year the Federal Statistics Office compute a standard hours of work for full-time workers, after allowing for holidays, sickness and trends in working hours.[1][2] For the years 1986 to 1988 this standard value has been 1620 hours per year. Thus the number of full-time equivalent workers in the industrial associations is 23 862.8 million workers [38 665.77 ÷ 1.620]. This calculation also applied to the Independent Insurance Association total hours worked. For 1988 the number of full-time workers in agriculture is estimated to be 1 873 641.[1][4] Multiplying this figure by 1620 hours gives an estimate of 3035.30 millions of hours worked in agriculture.

4   Another way of estimating hours of work in agriculture is to scale the standard yearly hours (1620) according to the ratio of agricultural to industrial hours per week – as given by the Labour Force Survey. The 1986 LFS found that people in employment (employees and self-employed) in industry worked 39.4 hours on average per week. The figure for agriculture is 51.1 hours, some 29·7% more. Making the tenuous assumption that the number of working weeks is about the same for agriculture and industry, the number of yearly hours is 2100 on average per agricultural worker, and the total number of yearly hours becomes 8381.5 millions (2100 hours x 3.9912 million workers).

5   The two estimates of yearly agricultural hours are very different probably because the high number of hours given by the LFS, a spring survey, may not be applicable throughout the year. In 1988, there were 278 non-traffic fatalities in agriculture. From this figure the frequency rate based on the conversion of full-time worker numbers is 0.092, and is 0.033 based on the average weekly hours in the LFS. Despite the weaknesses of both methods, the order of industries by frequency rate conforms to the order rate implied by the incidence rates given in Table 3 on page 52.

TABLE A4 Hours worked and frequency rates of fatal injuries F R Germany 1988

Industry	Hours worked (millions) (a)	Fatal injuries (b)	Frequency rate (c)
Energy	475.53	77	0.162
Manufacture	17 288.29	278	0.016
Construction	3 094.87	221	0.071
All services:	17 807.08	203	0.011
– transport	1 465.28	84	0.057
– other	16 341.80	119	0.007
All in industrial associations	38 665.77	779	0.020
Agriculture (d)	3 035.30	278	0.092
	8 381.50	278	0.033
Total	41 701.07	1057	0.025
	47 047.27	1057	0.022

Notes:
(a) From the *Commercial and Results Report 1988* of the Industrial Accident Insurance Associations (Ubersichtuber die Geschaftsund Rechnungsergebnisse der gewerblichen BG im jahre 1988)
(b) Fatal injuries excluding commuting and road traffic accidents
(c) Rate of fatal injury per million hours worked
(d) For agriculture two estimates of the hours worked are used to derive the frequency rate. The 3,035.3 estimate is derived from converting the official full-time worker total to hours worked. The higher estimate of hours worked is derived from the *1986 International Labour Force Survey* estimate of hours worked in agriculture

Frequency rates

6   Fatal injuries, hours of work and frequency rates are given in Table A4 for all main industry groups. Injuries and hours for workers acting in a voluntary capacity are included in the table.

Great Britain

7   Information on hours of work in Great Britain is taken from the *Labour Force Survey*. The survey provides the number of usual hours worked per week per employee but not the number of working weeks per year. However, it is likely that most employees work between 41 and 47 weeks per year – so that the holiday weeks are between 5 and 11 weeks per year.

8   From the *1988 Labour Force Survey* the average number of hours, including all paid and unpaid overtime is given in Table A5 for main

TABLE A5 Weekly usual hours worked on average by employees in Great Britain

Industry (a)	Hours (c)
Agriculture	42.3
Energy	42.5
Manufacturing	42.1
Construction	42.8
Services (b)	35.3
All (b)	37.6

Notes:
(a) Industry classified according to Standard Industrial Classification 1980
(b) Services and all industries includes the public sector
(c) Usual weekly hours, including all paid and unpaid overtime
Source – *1988 British Labour Force Survey*

TABLE A6 Frequency rates of fatal injuries for Great Britain 1987/88 and 1988/89 hours of work based on 41 and 47 working weeks per year

Industry (a)	1987/88		1988/89	
	Hours of work based on: (c)		Hours of work based on: (c)	
	41 Weeks	47 Weeks	41 Weeks	47 Weeks
Agriculture	0.039	0.034	0.040	0.035
Energy	0.038	0.034	0.025	0.214
Manufacture	0.011	0.010	0.010	0.009
Construction	0.059	0.051	0.056	0.049
Services (b)	0.002	0.002	0.003	0.003
All Above (b)	0.012	0.011	0.018	0.016

Notes:
(a) Industry classified according to Standard Industrial Classification 1980
(b) Fatals in services and all industries excludes public administration and education
(c) Hours of work are equal to the product of average weekly hours, number of working weeks and the number of employees

industry groups. Taking these average hours and multiplying by 41 and 47 weeks will give a range for the numbers of hours worked per year. The corresponding range of frequency rates are given in Table A6 for each main industry.

Bibliography

1   *The Statutory Accident Insurance in the F R of Germany in year 1988 (Die gesetzliche Unfallversicherung in der Bundes Republik Deutschland – im jahre 1988)* Statistical and Financial Report, Ministry of Labour, Bonn 1989

2   Private communication with Herr R Opfermann of the Federal Ministry of Labour, May 23 1990.

3   *Occupational Accidents and Diseases: a review of data sources* consolidated report, European Foundation for the Improvement of Living and Working Conditions: Dublin 1986 ISBN 92 825 6426 6

4   *Accident Prevention Report 1988: Report of the Federal German Parliament on the state of accident prevention and accidental incidents in the Federal Republic of Germany (Bericht der Bundesregierung uber den Stand der Unfallverhutung und das unfallgeschehen in der Bundesrepublik Deutschland Unfallvehetung-soericht 1988)* German Government: Ministry of Labour Document 11/5898

5   Private communication from Herr Dr B Hoffman, Industrial Insurance Associations, 18 May 1990

6   Private communication from Herr Dr I Brubach, Agricultural Insurance Associations, 21 May 1990

7   *Summary on the Commercial and Calculated new Results of the Industrial Accident Insurance Associations in year 1988 (Ubersicht uber die Geschafts und Rechnungsergebnisse – der gewer blichen Berufsgenossenschaften im Jahre 1988)* Hauptverband der gewerblichen Berufsgenossenshaften Sankt Augustin

8   *Labour Force Survey: Results 1986* Luxembourg, European Communities Commission, ISBN 92 825 8408 9

9   *Health and Safety Commission/Executive Annual Reports: 1987/88* ISBN 0 11 885476 3, *1988/89* ISBN 0 11 885531 X HMSO, London

10   *Yearbook of the International Labour Organisation 1988* Geneva

REPORT ON THE SYSTEM FOR THE HEALTH AND SAFETY PROTECTION OF WORKERS IN ITALY

By Lindsay Jackson and Rosie Edwards,
Health and Safety Executive

Introduction

1   This report was written by officials of the Health and Safety Executive in September 1990.

2   In Italy the responsibility for the prevention of accidents and ill health at work passed from the Labour Ministry to the Health Ministry in 1978; but the Labour Ministry still retains certain functions, including the representational role in Europe. The result is a system which is particularly difficult to co-ordinate; and which the Italians themselves have recognised as in need of reform.

3   This report concentrates on describing the Italian system at the time of writing, but also refers to recent suggestions for reform made by a senate parliamentary inquiry into conditions at work in August 1989.

Legal and institutional framework

General features of the system

4   Italy's law concerning health and safety at work derives from a number of different sources of varying levels of importance. The Italian Constitution of 1948 establishes the right of workers to health and safety protection. The Civil Code of 1942 requires employers to adopt suitable measures, in the light of present experience and technology, to protect the physical and mental welfare of workers. A series of particular decrees have also been passed on specific health and safety issues. Non-compliance with the law can lead to both civil and criminal liabilities.

5   Accident insurance is an important feature of the system. Employers are required to be insured by a government agency called the National Employment Injuries Institute – Istituto Nazionale per l'Assicurazione contro gli Infortuni sul Lavoro (INAIL) – which compensates workers for accidents at work unless the employer is found to be criminally negligent. Employers are required to report accidents at work to INAIL; and there is provision for employers to pay higher premiums if their safety measures have been ineffective.

6   Primary responsibility for workplace health and safety lies with the Health Ministry. Italy's national health organisation is based on local health units which are based on the territory of the commune and are intended to provide the full range of integrated health services, including hospitals, clinics, and the various prevention services including environmental protection and workplace health and safety. However, the regional automony of the Italian governmental system has had the result that the arrangements for health and safety protection at work vary very considerably from one part of the country to another. Although in 1978 central legislation was passed reforming health and safety legislation, regions had to promulgate their own regional legislation and were left considerable discretion as to how they carried the reforms into effect. Regional legislation varied. Some regions in the centre-north took the opportunity to do more than was required, but others did nothing. Few set up any regional office to co-ordinate the work at local level, or to prepare priorities. The exact role of the regions in the National Health Service and in workplace health and safety is acknowledged to be unclear. They have no formal duties or powers to direct the work of local health units or co-ordinate their work unless they have passed regional legislation to give them some powers in this sphere.

7   The local health units are autonomous bodies which can set their own priorities and methods of working. There is no formal means of liaison between local health units in the same region, and no central guidance on how local health units should relate to the region. The main control the Health Ministry has over the regions and the local health units is via the budget. There are some 660 local health units, but they are not spread uniformly over the country. They function less well in the south, where few have set up services for the prevention of accidents and ill health at the workplace. Where they have been set up, they are underfunded and understaffed. It has been estimated that resources for prevention services amount to 3% of the health service budget.

8   Before 1978, health and safety at work was the sole responsibility of the Labour Inspectorate, which is within the Ministry of Labour. In 1978, in a major reform, the National Health Service was set up, with the objective of unifying measures for prevention, rehabilitation and cure. Prevention of accidents and ill health was considered part of this and hence responsibility for prevention of accidents and ill health was given to the local health units. However the Labour Inspectorate retained a health and safety role and had the power to investigate accidents. In 1982, accident investigations were transferred to the prevention services of the local health units, but the Labour Inspectorate still undertakes this work, especially in areas of the country where the prevention services of the local health unit are not functioning. The Labour Inspectorate is responsible for aspects of nuclear safety, radiation

protection and enforcement of labour contracts, among other things.

9   The Health Ministry has primary responsibility for prevention of accidents and ill health at the workplace. A unit was set up within the Health Ministry to co-ordinate the work of the local health units and consult with the local health units and the regions; but further legislation to clarify the roles of the Ministry and the regions has not been promulgated. The legislation requires that a national health plan be produced; but it has not been used to date to set national priorities for action or general objectives concerning workplace health and safety.

10   Research and standards works is carried out by a government-funded institute, Instituto Superiore per la Prevenzione e la Sicurezza del Lavoro (ISPESL), which also undertakes certification and testing duties and gives expert advice to the inspectors of the local health units.

11   At the level of the firm, collective agreements form an important source of control over workplace health and safety. These do not have the power of a statute, but are nevertheless upheld by the courts as binding contracts. Workers have the right to organise workers' councils. Recently proposals have been made for the setting up of joint safety committees and the creation of safety delegates. One problem in Italy is the very high proportion of small firms (defined as employing 15 or less); workers' councils are generally found in larger firms with a trade union presence. The recent proposals in the Lama Report (see paragraph 21) suggest the creation of safety delegates representing workers in several firms in a geographical area in a particular sector.

12   The 1978 reform requires that firms set up workplace and occupational hygiene and medical services. These exist only in major firms in the centre-north. The law also empowers local health units to intervene and provide such services themselves if firms fail to do so. One of the proposals in the Lama Report (paragraph 21) is to provide clear definitions of categories of firms which have to provide such a service.

General principles of the law

13   As was explained in paragraph 4, Italian law derives from a number of sources. Both labour law and social security law are relevant.

14   The most important general principles are the duty of the employer to do all that is technically feasible to achieve conditions which are healthy and safe and the right of workers to take direct responsibility for their own health and safety.

15   There are many specific decrees imposing requirements on certain sectors. Guidelines are issued by the Health Ministry, often on the advice of the research institute ISPESL, to keep up with new working procedures and new risks. These can be enforced by local health units and failure to comply amounts to a breach of the law.

16   Accident investigations are undertaken by magistrates, using inspectors either from the local health units or the Labour Inspectorate to do this, and with the power to instigate criminal proceedings in serious cases. Also the inspectors of the local health units have the power to issue the equivalent of enforcement and prohibition notices. Inspectors can in addition go to a magistrate to institute proceedings in the case of an employer's contravention of health and safety law.

17   Social law is also relevant, because as was stated in paragraph 5, employers have to be compulsorily insured and are thereby exempted from civil liability for work accidents unless found to be criminally responsible for the act giving rise to the accident.

Inspection arrangements

18   As was explained in paragraphs 6 and 7, the local health units cover the full range of health services and are autonomous. The staff of the prevention services of the local health units are engaged by open competition at regional level. In 1989, it was estimated that the prevention services of the local health units had 3000 professional staff, but these are unevenly spread around the country. There is no national inspection plan, but some regions and some individual local health units have attempted to prioritise inspection and organise planned interventions in particular sectors of industry. There are regional variations in the way prevention services are set up. In some cases, prevention of health and safety at the workplace is carried out by non-specialist units responsible for public health generally. In other cases, prevention is carried out by one local health unit prevention service on behalf of several local health units.

19   In addition, there are 1744 inspectors in the Labour Ministry involved in enforcing labour regulations generally; health and safety forms a small part of this work.

Occupational health services

20 The Labour Inspectorate includes a number of medical inspectors, based both in central and regional offices, who co-operate with the local health units as appropriate. Regular medical examination of the workforce is required by a works doctor. The prevention services of the local health units ensure that medical examinations are carried out. They also have powers to intervene if a firm fails to set up its own occupational hygiene and medical services (see paragraph 12), but this right has never been exercised.

Proposals for reform

21 Proposals for reform have recently been made by a senate parliamentary inquiry into conditions at work – President, Senator Lama – which reported in August 1989 (the Lama Report). The Lama Report proposes that the Health Ministry should be confirmed as having the primary responsibility for health and safety at the workplace, but that its powers and duties should be made more specific. The Report proposes that the Labour Inspectorate should be expanded but should lose all its health and safety functions; that the staff of the local health units should increase, over a three year period, to 6000 and ultimately to 12 000; that national and regional plans and priorities should be established; and that the Health Ministry should take over from the Labour Ministry the duty of representing Italy on international bodies dealing with health and safety at work. In addition, the Report proposes new legislation which would codify existing law; establish workers rights to appoint safety delegates; and implement EC directives with modifications to suit the Italian system.

Bibliography

1 *Comparative analysis of the reports of tripartite missions assessing the effectiveness of labour inspection systems in seven countries of Western Europe* International Labour Office, Geneva, 1985, ISBN 92 210 5242 7

2 *The law and practice concerning occupational health in the member states of the European Community* prepared by Environmental Resources Ltd for the Commission of the European Communities, vol 4, 1985

3 *Labour inspection in the European Community* Stewart Campbell, Health and Safety Executive, HMSO 1986, ISBN 0 11 883871 7

4 *Handbook of labour inspection (health and safety) in the European Community* D E Clubley, Health and Safety Executive internal report 1990

5 *Senate Parliamentary Inquiry into Conditions of Work in Italy, President, Senator Lama:* report, August 1989

NATIONAL OCCUPATIONAL
ACCIDENT STATISTICS: ITALY

by Graham Stevens,
HSE's Statistical Services Unit

Introduction

1   This note briefly describes the system of reporting occupational accidents in Italy and makes a crude comparison between Italy and Great Britain in terms of reported injuries in 1983 and 1984.

2   Occupational accidents are reported to the main National Institute for Insurance against Accidents at Work – Istituto Nazionale per l'Assicurazione contro gli Infortuni sul Lavoro (INAIL). Employers and the self-employed pay premiums to INAIL who compensate accident victims. INAIL publishes the occupational injury statistics for Italy. Many non-manual workers are not insured with INAIL, and therefore their injuries do not feature in the statistics. Nevertheless, the Italian fatal injury incidence rates derived in this note are substantially higher than those of Great Britain. Furthermore, this comparison is not likely to be greatly affected by over-reporting of injuries because there are checks on, and no incentive for, false claims related to work injury.

Insurance

3   Employees and the self-employed are insured against accidents at work by INAIL. Employers send accident reports to them to assess claims for compensation. INAIL publishes statistics on occupational injury (and ill health) for Italy.[1][3]

4   There are some notable exceptions to the coverage of workers by INAIL. Firstly, non-manual workers in the public sector are not covered. Though postal and transport workers in the public sector are covered.[3] Secondly, in general, non-manual workers are not covered unless they are working with certain electrical equipment – though the extent of this exclusion is far from clear.[2]

5   Employers, including subcontractors, are required by law[2] to register with INAIL. Self-employed people and family workers in agriculture must also be registered. Therefore, most agricultural workers are covered.

6   INAIL pays compensation to work accident victims depending on the degree of severity of injury.[2] Employers pay annual premiums to INAIL. This payment is assessed by INAIL depending on the industry activity and the safety record of the employer.[2] There is thus an incentive for employers not to collude with false claims. Additionally, the payment of compensation and medical expenses is an incentive for employees to ensure the reporting of accidents. Under-reporting is reckoned to be negligible.[3]

7   There is the special case of heart attacks causing deaths at work and hence over-reporting. These are reckoned[2] not to be reported because of the system of determination of premiums mentioned earlier and also because INAIL will examine the medical history of the victim and the reasons for death.[2] If an injured worker dies much later, any claims by relatives must prove to INAIL the work causation.

The reporting system

8   Employers and the self-employed must complete a report of an occupational injury for INAIL within 24 hours of the accident. INAIL usually send a copy of the report to the Labour Inspectorate. If an injured worker dies in hospital soon after an accident, the hospital notifies the police. They notify the Labour Inspectorate who inform INAIL.

9   Reportable accidents are those accidents at work to insured workers which makes them absent from work for more than three days (not counting the day of accident and including weekends). Commuting accidents are generally not included in this. However, road accidents occurring in the course of work are included. The main point is that this non-fatal accident category in Italy is defined by absence in much the same way as the over-3-day category is in Great Britain.

10   These over-3-day injuries are classified by INAIL into one of three severities of disability. The most severe is death, followed by permanent disability which reduces the ability to work by more than 11%. The least severe is the temporary injury with either no permanent effects or permanent effects of less than 11% disability. The permanent disability injury is in no way comparable with the British definition of a major injury.

Injury statistics

11   INAIL publishes injury statistics for Italy based on those reports it receives. Notiziario publications,[1] while publishing frequency rates of injury, do not give the breakdown of absolute numbers of injury by the three disability severities. Neither does INAIL publish employment or associated injury incidence rates. However, INAIL will have supplied the International Labour Organsiation (ILO) with injury statistics for their 1988 yearbook.[4] No employment figures were given for this edition.

12   Table 1 displays reported injuries in Italy as published by the *1988 ILO Yearbook*. The figures are for 1983 and 1984. The fatal injury figure for 1984 is consistent with the fatal frequency rate and hours of insured workers published by INAIL.[1]

TABLE 1 Reported (a) occupational injuries (b) in Italy 1983 and 1984

Year	Fatal	All reported
1983	1 587	876 324
1984	1 579	866 539

Notes:
(a) Source – *International Labour Organisation Yearbook 1988*
(b) Excludes commuting injuries but includes road injuries occurring in the course of work

13   While no absolute figures relating to employment or insured population are given to the INAIL publications, they can be obtained by the European *Labour Force Survey*.[5] In the 1986 Survey, there were 20 684 000 in employment in Italy. Of these 14 682 000 were employees, 4 935 000 were self-employed, and 1 067 000 were family workers. Agriculture accounted for 23% of the latter two groups and 10·5% of all employment.

14   With these employment figures, rates of injury can be derived that understate the true rate in Italy because of injuries to non-manual workers excluded from insurance, and hence from the figures. Table 2 displays the resulting injury incidence rates and also frequency rates based on the ILO figures for injuries and the hours of work given by INAIL. The frequency rates are, in effect, rates of injury for mainly manual labour.

TABLE 2 Injury incidence and frequency rates in Italy 1983 and 1984

Year	Injury incidence rate (a)		Frequency rate (b)	
	Fatal	All reported	Fatal	All reported
1983	7.67	4 237	0.073	40.1
1984	7.63	4 189	0.072	39.6

Notes:
(a) Rate expressed per 100 000 in employment. Employment figure for 1986 from the *1986 International Labour Force Survey*
(b) Rate expressed per million hours worked in a year by the population insured with INAIL

15   The industrial classification is not readily comparable with that of Great Britian. For example, fishing, forestry and animal rearing is part of the manufacturing activity devoted to food manufacture. This group contains, what in Great Britain would form part of agriculture and part of manufacturing. Also, mineral extraction and coal mining, separate in British sectors, are in one Italian sector. There is little point in producing injury rates by Italian main sectors because these would not be comparable with British industrial sectors. Furthermore, the injury figures of Table 1

include road accidents occurring in the course of work.

Comparisons between Italy and Great Britain

16   The comparison between Italy and Great Britain in this note is necessarily based on the injuries reported for all industries combined and published by ILO. These injuries exclude many non-manual workers as intimated by the description of the insurance coverage. Therefore the injury incidence rates, derived from employment estimates covering all sectors, will understate the true level of injuries. The amount of understatement is difficult to assess but might not be substantial if, as in Great Britain and France, the bulk of fatalities occur to manual or operative workers.

17   Another important factor in the comparison is the inclusion in Italian figures of road accidents occurring in the course of work. Such accidents, judging from French, German and Spanish figures, are only around 1 to 2% of all-reported injuries but could be 30 to 40% of published work fatalities. The lower limit of this range would represent accidents not likely to be reportable in Great Britain, while the upper limit (from French fatals) will include some reportable deaths. Subtracting 40% from the Italian fatal rate would over adjust for road deaths. When between 30% and 40% is subtracted from the 1984 fatal incidence rate, the result is a rate exclusive of road accidents of between 4.58 and 5.34.

18   The focus of the argument is that, while direct comparison cannot be made, this note can derive a lower limit for the fatal rate in Italy – a limit that is substantially above fatal injury rates in Great Britain. Over-reporting of fatal injuries is not likely to affect such a comparison. Fatal injury rates for Italy and Great Britain are displayed in Table 3 for the years 1983 and 1984, together with an estimated lower limit for Italian

TABLE 3 Fatal injury rates for Italy and Great Britain: all industries 1983 and 1984

Year	Italy (a)		GB (b)
	Rate	Rate excluding work road deaths	Rate
1983	7.67	4.60	2.2
1984	7.63	4.58	2.1

Notes:
(a) Fatal injury incidence rates expressed per 100 000 in employment (employees and self-employed people). Rate based on deaths mainly to manual workers and operatives, including some from the Public Sector. Employment is from the *1986 International Labour Force Survey*
(b) Fatal injury incidence rate expressed per 100 000 employees. Rate is based on all reported deaths including the public sector

deaths exclusive of road accidents occurring in the course of work. Fatal injury rates in Great Britain are substantially lower than Italian rates even with an allowance for road accidents.

19   The comparison must consider the different industrial mix in the two member states. From the *1986 Labour Force Survey* (LFS), around a third of employment is in the manufacturing and construction industries in Italy and in Great Britain. However, agriculture accounts for 10·5% of employment in Italy and only 2·2% in Great Britain. The service industries in Italy are almost correspondingly smaller than in Great Britain – in percentage terms. While no sectoral rates are obtainable for Italy, the British rates can be adjusted to allow for the Italian industrial mix. The British fatal rate for 1983 is 2.2 from Table 3. This rate, adjusted for Italian industrial mix as given by the *1986 Labour Force Survey*, is 3.0. So, depending on road injury numbers, between over one sixth and a third of the difference in the fatal rates for 1983 can be accounted for by differences in industrial mix (at a broad level).

20   Finally, we should consider the comparison in terms of frequency rates of fatal injury – per million hours worked per year. The hours of work and frequency rates given by INAIL[11] are related to mainly manual workers. These rates are thus expected to be higher than if calculated for the working population including non-manual labour.

21   A frequency rate can be determined for manual/production workers in Great Britain. Occupation of the injured is recorded by injury reports made to HSE's factory and agricultural inspectorates only. In 1988/89 some 40 out of 289 fatal reports were either to managerial, clerical or education personnel. Most of the fatal reports to the other inspectorates do not involve non-manual personnel. The 1988/89 fatalities include 167 in the Piper Alpha tragedy. Subtraction of the 40 cases gives a slight over estimate of the total number of fatalities to manual/operative personnel (490).

22   A survey[6] on British hours of work estimated that manual labour in production industries worked 43.6 hours per week on average in 1988. This is higher than the average weekly hours for all employees as given by the *Labour Force Survey*. But this is consistent with the high number of hours worked by employees in production industries found by the *Labour Force Surveys* of 1986 to 1988. In order to derive a frequency rate, we must estimate the number of manual/operative personnel. Such personnel account for nearly 53% of employees in the *1988 Labour Force Survey*. Applying this to the Employment Department

estimate of employees in 1988/89 gives 11 529 650 manual employees. Taking a 41 week working year these manual employees worked 20 610.4 million hours in 1988. The corresponding frequency rate is 0.024. The frequency rate assuming a longer working year of 47 weeks is 0.021.

23   The derived British frequency rates are for employees who are not in professional, clerical, or managerial occupations. The frequency rates for Italy are based on insured people who are in manual occupations, 'work with machines', and are not supervisors of manual labour. These frequency rates include some workers from the public sector (transport, postal workers and drivers). The coverage of sectors is marginally different in the two countries in that some distribution employees, counted in Italian rates, are not included in the working hours used in the derivation of British frequency rates. Also, some supervisors in the services will be counted in hours worked for Great Britain but not for Italy. Nevertheless, the fatal frequency rate for Great Britian in 1988/89 is much lower than the Italian frequency rates for 1983 and 1984. The frequency rates in Great Britain are only likely to be marginally higher in 1984 in keeping with the slow decline in fatal injuries since then.

Conclusion

24   In terms of both the incidence and frequency rate of fatal injury, Great Britain is lower than Italy. Over-reporting of fatalities in Italy and industrial mix are not important factors here. The comparison is hindered by the different coverage of types of worker, manual or not, in the two countries and the inclusion of work related traffic deaths in Italy. Allowances for these points still leave Great Britian much lower in rates of fatal injury than Italy. This suggests that the risk of a work fatal injury in Great Britain is less than in Italy.

Bibliography

1   *Notiziario Statistico 1986* Volume 1 – 3. Istituto Nazionale per l'Assicurazione contro gli Infortuni sul Lavoro

2   Private communication 13 July 1990 Signor Enrico Martoni of Ministero del Lavoro, Roma

3   Interim Report by Roger Clark, July 1990. Report submitted to CEC Working Party on Occupational Accident Statistics, 18 July 1990

4   *Yearbook of the International Labour Organisation* 1988 (Table 27) Geneva

5   *Labour Force Survey: Results 1986* Luxembourg, European Communities Commission, 1988, ISBN 92 825 8408 9

6   Earnings and hours of manual employees in October 1989; *Employment Gazette,* May 1990, vol 98, no 5, 244-254

REPORT ON THE SYSTEM FOR THE HEALTH AND SAFETY PROTECTION OF WORKERS IN SPAIN

By Sally Van Noorden and Kevin Myers, Health and Safety Executive

Introduction

1   This report was prepared by officials of the Health and Safety Executive in November 1990. It is based on information gathered in September 1990; at which time a draft law codifying the Spanish provisions on health and safety had been published for comments prior to being laid before the Spanish Parliament. This draft was intended to meet criticisms that the current provisions were too disperse and outdated.

2   The new draft law is designed to bring Spain into line with EC legislation, in particular the Framework Directive. To further this end, it contains provisions bringing public servants under health and safety at work legislation (previously they were not covered). The law also creates a tripartite Higher Council to advise the Government on health and safety measures. This council would advise both the Ministry of Labour and the Ministry of Health. (There is already a tripartite governing body overseeing the National Institute for Health and Safety at Work, which is the research arm of Spain's health and safety system.)

3   New general legal requirements are proposed as follows:

    (a) employers are to be given a general responsibility to avoid risks to workers' health and ensure that they carry out their work in safe conditions 'so far as is reasonable and possible'; this formula is said to have been derived from an International Labour Organisation (ILO) convention;

    (b) workers are given the right to be informed of periodic assessments that the employer must make; the right to health surveillance; and the right to stop work if they think they are in immediate danger. (This last right does not formally exist in the current law, although it has evolved over the years as a result of judicial rulings);

    (c) employees are required to co-operate with employers to enable them to fulfil their obligations.

4   Finally, the new draft law envisages a wider occupational health service than exists in Spain at present, in that it requires all companies to have access to preventive services and creates new committees of 'health delegates' at the level of the firm.

5   The draft will be amended in the light of comments from the social partners and others.

Legal and institutional framework

General features of the system

6   Spain's law concerning health and safety at work derives from a number of different sources of varying levels of importance. The Constitution of 1978 requires the public authorities to safeguard workers' safety and health. International obligations assumed by Spain such as ILO conventions are taken to be part of domestic law. The Spanish Parliament has passed a number of specific acts dealing with health and safety, in particular the Workers' Statute of 1980 which states that workers have a general right to safety rules and a responsibility to observe the rules adopted by the employer. There are also extensive administrative regulations establishing the Labour Inspectorate and laying down precise and technical health and safety rules. These regulations can be issued either by ministers or by officials.

7   Accident insurance is an important feature of the system. Employers can seek insurance cover either from a government agency, the National Institute of Social Security, or from associations made up of member companies called Mutuas Patronales. These are effectively agencies authorised by the Ministry of Labour and Social Security and operate on a non-profit basis. They are funded by employers' contributions channelled through the Ministry. Employers are required to report accidents at work to the insurance authorities, who pass the information on to the Labour Inspectorate.

8   The primary role of the accident insurance associations is to manage compensation. They do have a preventive role, but this is a secondary and (in terms of resource allocation) minor function. There is one important exception to this, namely the Government Institute for the Protection of the Health of Workers at Sea.

9   The Spanish Labour Inspectorate is part of the Ministry of Labour and Social Security. It is responsible for the implementation of a wide range of legislation including social security, employment and industrial relations issues, as well as health and safety at work. It concentrates its efforts on employed people within the workplace; protection of the public from work activities is the responsibility of other bodies. The Inspectorate is led by a director general who heads a unit in the Ministry. Although the Labour Inspectorate is a single national inspectorate controlled from the centre in Madrid, there is in each province a director of the Ministry of Labour

and Social Security who is responsible in that province for the activities of the Ministry. In certain provinces, called the 'autonomous communities'*, the responsibility for labour and industry has been devolved from central government to the local provincial authority. In each province there are local commissions chaired by a provincial director for labour and social security and consisting of the Labour Inspectorate, the local authority, trades unions and employers. Notwithstanding the apparently devolved nature of the system, the national inspectorate is directed centrally and a national programme of work is set by the director general.

10   Specialist health and safety advice to the Labour Inspectorate is provided by the National Institute of Health and Safety at Work (Instituto Nacional de Seguridad e Higiene en el Trabajo, INSHT), which is funded by the Ministry of Labour. INSHT employs civil, mechanical and electrical engineers, occupational hygienists and doctors who provide advice and information to industry and commerce, and assist labour inspectors on request and at accident investigations. They have no inspection or enforcement function. An important part of the work of INSHT is the production and publication of health and safety information in the form of leaflets, posters and pamphlets. They are also involved in the training of safety specialists, employees and union representatives.

11   The Institute is also responsible for carrying out a national survey on conditions of work. This consists of a compilation and analysis of information culled from a national questionnaire on workers' conditions which is sent out to employers and employees every five years. It also carries out sectoral surveys between the quinquennial national surveys. This aspect of their work is, in their view, a requirement of ILO conventions.

12   There is a Spanish standards institute sponsored by the Ministry of Industry; but in practice, technical work on safety standards and representation in Europe is carried out by staff of INSHT.

13   At the level of the firm, regulations provide that larger companies must have committees on safety and hygiene, consisting of safety experts and representatives of employers and workers. The largest companies (those with over 100 people) are required to have medical personnel whose performance is monitored by the Ministry of Health. Smaller companies are expected to have safety officers.

General principles of the law

14   As was explained in paragraph 6, Spanish law currently derives from a number of sources. Both labour law and social security law are relevant.

15   The detail of the law is found in regulations. The most important of the regulations concerning health and safety at work is the General Ordinance Concerning Safety and Health at Work of 1971 (Ordenanza General de Seguridad e Higiene en el Trabajo). This sets out the powers of the Ministry of Labour to make rules, institute research and give technical guidance; lays down the functions of the Labour Inspectorate; establishes provincial occupational safety and health councils; lays general duties on both employers and workers; sets out the functions of occupational safety and health committees and of safety officers; and makes provision for administrative fines.

16   At the same time there are numerous other regulations which set out precise technical standards, eg on protection against lead, asbestos and ionising radiations.

17   The law makes provision for administrative fines. These are divided into three categories: 'light'; 'serious'; and 'very serious'. These fines apply to contraventions of wages and employment law as well as health and safety legislation. The appeal process is largely administrative, but further appeals can be made to the courts once the administrative route has been exhausted.

18   There is provision in the Spanish criminal code for criminal sanctions in the case of incidents or accidents that have occurred because of health and safety at work violations, but these are rare. Mostly, health and safety violations are administratively sanctionable (the theory being that the criminal provisions are reserved for acts of commission, administrative fines for acts of omission). Double sanctions of the same act are prohibited; if criminal proceedings are being undertaken, administrative sanctions may not be imposed. In addition, the criminal judge may determine the extent of private civil liability, for eg injuries caused.

19   Social security law is also relevant. Insurance against accidents and ill health at work is treated separately from the rest of the Spanish social

*Only 7 of the 17 'autonomous communities' have devolved health and safety powers: these are Galicia, Basque, Andalucia, Canaries, Novarra, Catalonia and Valencia.

security system in that contributions are paid only by employers and compensation is more generous than for illnesses and accidents not occurring at work. The payment of contributions and the disbursement of compensation is managed by the Mutuas Patronales, which are the employers' insurance associations. An exception to this is the insurance scheme for maritime workers, which is managed by the Ministry. Where work accidents or occupational illnesses are due, in the opinion of a labour inspector, to lack of health and safety measures having been taken, the employer is required to pay an extra surcharge to the employee of some 30% of the compensation. The employer is not reimbursed by the insurance organisations for this surcharge. The law provides that the following circumstances constitute non-observance of health and safety measures:

(a) machines and installations which do not have the required safety devices;

(b) failure to use or maintain such devices;

(c) non-observance of general or particular measures for health and safety at work;

(d) non-observance of elementary health rules; and

(e) unsuitability of the worker for the work position, with the worker's characteristics, age, sex and other conditions having been taken into consideration.

Inspection arrangements

20   The Spanish Labour Inspectorate is a generalist one, covering a wide range of legislation. It is estimated that about 60% of the work of the Labour Inspectorate is spent on monitoring social security contributions; 25 to 30% on employment and working conditions; and 15 to 20% on health and safety issues.

21   The Labour Inspectorate has an important role in arbitrating in disagreements and disputes between employers and employees, particularly in situations where employees are working in dangerous conditions, and negotiate for the payment of a bonus. Although the policy of the Labour Inspectorate is to endeavour to promote negotiations in such a way that the danger-money element is removed, they have to arbitrate in particular cases. If, in the course of their duties, inspectors need a technical assessment of conditions such as noise, heat or risk, the advice of the National Institute of Health and Safety at Work is sought.

Occupational health services

22   Spain does not have a separate medical inspectorate. Companies are required to have access to appropriate medical personnel; the largest companies (those with over 100 employees) have their own, while the smaller ones can call on health and safety consultancy services and health care facilities provided by the accident insurance associations. Under the new draft health and safety law, currently out for consultation, an expansion of occupational health services is envisaged for the future, but the mechanics of implementing this have yet to be worked out. The Government's intention is that the Ministry of Health should play a large role in the development of occupational health services.

The duties of inspectors

23   The Labour Inspectorate produces plans nationally, and monitors work activity from the centre in consultation with chief inspectors in provincial offices; although in those provinces which are autonomous communities, the head of the local government can exercise an important controlling interest.

24   There are two grades of inspector: labour inspectors and labour controllers. Labour inspectors are university graduates who can inspect all sizes of premises and carry out the full range of inspection duties. Labour controllers are usually non-graduates; they inspect only premises employing less than 25 people and only for employment and social security issues, not health and safety. Admission to the Labour Inspectorate is by written and oral competition. In 1990, there were some 500 labour inspectors and 700 controllers; this cadre is set to expand to cope with the obligations laid down in the new health and safety law.

25   Inspectors have powers of entry to premises (except in the case of private domestic establishments), but many visits are made by appointment in response to requests for advice. A questionnaire approach is often used for planned inspections.

26   Considerable use is made of administrative fines; about 12 500 are imposed annually. About one third of these fines are appealed against (to a higher level in the Ministry); usually these appeals are not successful. In some of the autonomous communities in the south of the country, the local authorities have been given the right to determine sanctions questions rather than the Ministry in Madrid.

27   Labour inspectors are assisted by engineers, occupational hygienists and doctors from the National Institute of Health and Safety at Work who undertake measurement, recording and risk assessments. As the labour inspectors are generalists (mostly law graduates), the professional advisers from INSHT play an important role. INSHT has four national centres in different parts of the country undertaking specialised research in different areas. It also provides much of the representation in Europe on specialised health and safety questions, and on standards. Forty to fifty percent of INSHT's time is spent on assistance to the Labour Inspectorate, and the remainder on work on its own initiatives.

28   Provision is made for liaison between the Labour Inspectorate, trade unions, the provincial authorities and employers through local commissions. These meet regularly to discuss trends in accidents and occupational disease figures provided by the National Institute of Health and Safety at Work.

Bibliography

1   *Handbook of labour inspection (health and safety) in the European Community* D E Clubley, Health and Safety Executive internal report 1990

2   Account of the Spanish legal health and safety system by Dr Joaquin Aparicio Tovar in a report prepared for the Health and Safety Executive by the Institute of Advanced Legal Studies, July 1990

3   Text of the new Spanish law and information booklets, supplied by the Spanish Ministry of Labour and Social Security and the National Institute of Health and Safety at Work, September 1990

NATIONAL OCCUPATIONAL ACCIDENT STATISTICS: SPAIN

by Graham Stevens,
HSE's Statistical Services Unit

Introduction

1   This paper briefly describes the system of reporting occupational injuries in Spain and indicates how these compare with the injury record of Great Britain.

2   In summary, work injuries are reported to insurance bodies who make payments for associated lost work and for medical treatment. This would suggest that there is no serious over-reporting, though the various insurance bodies will monitor claims for such payments in different ways. A crude comparison of fatal injuries would suggest the rate, incidence or frequency, of fatal injury is less in Great Britain than in Spain.

Insurance

3   Employers must register by law with either the National Social Security System – Instituto Nacional de la Seguridad Social (INSS) or with one of the 100 or so private insurance organisations. These organisations also cover general insurance. Employees are given work injury insurance by these bodies to whom employers give annual premiums.[2] The self-employed are not covered, though agricultural self-employed are. Many public sector employees are not covered – though some manual workers are. The coverage is not tightly defined.[1][2][3]

4   Non registration by employers is probably minimal[2][3] and would be exposed following an accident leading to required medical treatment. Some larger companies can self-insure. Under-reporting is regarded as negligable.[2]

5   The insurance bodies reimburse employers for any wages paid to workers absent due to a work-related injury. They also pay expenses associated with medical treatment of an injured employee. Ordinary sick pay is payable to employees absent due to non-work accidents but is less than payments for work-related absence.[2] The employer's premiums to the insurance bodies is related to company size and industry activity.[2]

The reporting system

6   Following an injury to an employee, the employer must complete a five-part injury report form sending three copies to the appropriate insurance body,[2][3] giving one to the employee, and retaining a copy. The insurance body sends a copy to the Labour Inspectorate and the Ministry of Labour. The published statistics[3][4][5] are derived from this latter source. It is not clear what cross checks are made to see if all reports arrive at the Ministry of Labour.[3]

7   Employers are required to notify the insurance body within 48 hours[2][6] for all severities of injury. Under-reporting by employers is thought to be minimal,[2] and is probably because employees are paid for days lost by the insurers.

8   On over-reporting, there is no incentive for employers or insurance organisations to allow false claims to go forward for payment. To be set against this is that (INSS) and the 100-plus private insurance organisations will assess claims in different ways. It is not clear how this is done nor how heart attacks at work are excluded from the figures on work fatalities.[2] However, in general non-work accidents do not feature in the occupational injury figures,[2] and duplicate reports on the same injury are excluded.[9]

Definition of reportable injuries

9   Any injury at or connected with work is reportable.[2][7] Those involving an absence from work are reported through the insurance systems described earlier. The main definition is the 'over-1-day' injury – an injury leading to more than 24 hours' absence not counting the day of injury or return to work. However some under-1-day absences also get included in the published figures.[1][3] Injuries involving no absence from work are reported through a separate system.[2] Work injuries include occupational road traffic and also commuting accidents. The latter are published separately.[4][5]

10   Occupational fatal injuries are those where the victims die, either on the same day as the accident or soon after.[1][2] While there are no figures for the numbers of work injuries with deaths delayed up to a year (the definition in Great Britain), the number of deaths that are not recorded are not likely to be many. The fatal rates for Spain produced in this paper therefore understate the true rates. This does not affect the comparison with Great Britain whose fatal rates are lower.

Some injury statistics for 1988

11   This section presents some employment and injury figures published by the Ministerio de Trabajo y Seguridad Social, (Ministry of Labour and Social Security)[6]. The employment figures given are the average of the quarterly surveys by the Spanish National Statistical Institute (Instituto Nacional Estadisticas – INE). Since the numbers of insured workers is not readily available from the private insurance organisations, the INE salaried employee estimates have been the best source of the exposed

population that give rise to the reported injuries.[5] In a recent change in methodology, the Statistics Department of the Spanish Ministry of Labour have derived rates of injury using the numbers of workers who are affiliated with the social security systems.[8] The numbers of affiliated workers so used are not very different from the survey estimates[9] and, from examination of the recently published injury rate for 1988,[8] represent the employee population with the addition of the self-employed in agriculture. The vast bulk of other self-employed are affiliated but are not covered by injury reporting requirements.

12   For the year 1988, the estimated number of people in employment from the INE surveys is 11 772 700. Of these, 8 351 500 are salaried employees. The numbers in employment by five main industry sectors are given in Table 1. These sectors consist of the same sorts of activities in Spain as in Great Britain. Details of the Spanish industrial classification are given in the appendix.

TABLE 1 Employment in Spain 1988 (000s)

Industry	Salaried employees	Self-employed and employers	Total
Agriculture	541.4	1 152.8	1 694.2
Energy	136.9	1.2	1 038.1
Manufacturing	2 321.1	345.9	2 665.8
Construction	774.2	246.1	1 020.3
Services	4 577.8	1 676.4	6 254.2
All industries	8 351.5	3 421.2	11 772.7

13   The numbers of affiliated workers for 1988, comprising employees and the self-employed, is 11 609 500 and is not far from the INE survey estimate of 11 772 700. Affiliated employees and the self-employed in agriculture form the exposed population covered by injury reporting regulations. Their numbers are 9 257 200: the corresponding INE survey estimate is 9 504 300. The INE estimates will contribute to injury rates in this paper because they are readily available by industry sector whereas the affiliated numbers are not. The effect on injury rates is to reduce them by 2½% over all industry.

14   Agriculture is the industry with the largest proportion of self-employed in total employment. As in Great Britain and West Germany, manufacturing and energy have the smallest proportions of self-employed people in total employment. The distribution of employment in Spain is more weighted towards the traditionally riskier manufacturing and construction industries than in Great Britain. For example in 1987 manufacturing accounted for 28·5% of

employees in Spain but 23·9% in Great Britain. The figures for construction were 8·7% and 4·7%. Further details are given in Table A2 of the appendix

15   Some injury statistics and rates are presented for the years 1987 and 1988. The source of these are published reports from the Ministry and Social Security.[4][8] In 1987, 530 946 work accidents involving time away from work were reported to the Ministry of Labour. These were based on copied reports from the insurance bodies as described earlier. They are based on employees with some self-employed in agriculture. There were another 139 056 reports of injury where no time loss was involved and an additional 40 254 'in transit' home to/from work injuries. Time loss injuries are categorised into one of three severities: slight, severe ('graves') and fatal. A change in administrative procedures (1988) has led to an increase in this number of reports which involve no time loss. They include some injuries involving less than 24 hour absences from work.[1][3]

Table 2 gives the numbers of injuries according to severity and also the incidence rates for time loss.

TABLE 2 Number of injuries in 1987 and 1988

Time loss	Severity	Number 1987	Number 1988
Some time loss	Sight	518 406	563 759
	Severe ('graves')	11 358	13 958
	Fatal	1 182	1 288
No time loss	All time loss	530 946	579 032
		139 056	442 324
In transit	All severities	40 254	36 016
All accidents	All severities	710 256	1 057 372

Comparisons between Spain and Great Britain

16   This section presents rates for all-reported and fatal injuries in 1987 and fatal rates for 1988. Employees are covered by insurance with the exception of some employees in the public sector.[3] Manual workers in the public sector are covered. Because of the public sector, any rate based on reported injuries will be a slight under estimate of the true rate. Further, some self-employed in agriculture are insured and so their injuries are reported. Rates for agriculture can be derived from INE survey estimates of all those in employment (employees and self-employed).

17   Another point is that road traffic injuries are treated as occupational by the insurers and so feature in the statistics. These could include lorry and bus drivers and employee passengers injured in the course of their work on the road. The 'cause' classification includes such injuries under a

TABLE 3 Injuries and rates for Spain 1987 and 1988

Industry 1987	All reported time loss (a)	Fatal injuries (d)	All reported rate (b)	Fatal rate (b)(d)
Agriculture	48 538	194	2 825	11.3
Energy (c)	18 421	59	13 930	44.5
Manufacturing	235 431	316	10 363	13.9
Construction	88 013	224	12 691	32.3
Services	140 493	389	3 234	9.0
All Industries with vehicle accidents	530 946	1 182	6 659	14.8
without vehicle accidents			6 565	9.7
1988				
Agriculture	45 384	149	2 679	8.8
Energy	19 613	63	14 327	46.0
Manufacture	255 900	261	11 025	11.2
Construction	103 865	234	13 416	30.2
Services	154 270	265	3 370	5.8
All industries	579 032	972	6 090	10.2

Notes:
(a) Reportable injuries mainly arise from over 24-hour absences from work (over-1-day) but will include some under-1-day accidents
(b) Incidence rate per 100 000 employees, except in agriculture where injuries are for expressed per 100 000 in employment
(c) Energy comprises: coal mines, petroleum extraction and refining, electricity, gas and water
(d) **Fatal injuries in 1988 exclude vehicle accidents for each industry sector**

category of 'being knocked down or hit by vehicles'. This 'vehicle' category will not include overturning vehicles such as fork-lift trucks but will include the reversing vehicle accidents that are reportable in Great Britain under the current reporting regulations. For the years 1986 to 1988 vehicle accidents account for around 1% to 2½% of all-reported accidents. In contrast they account for between 24% and 35% of fatal injuries. For comparison with Great Britain, vehicle accidents are excluded from the Spanish figures, particularly for fatal injuries. This exclusion will remove some Spanish injuries that would be reportable in Great Britain. However, the British fatal rate is below the resulting estimated Spanish fatal rate which in turn understates the Spanish true rate.

18   The distribution of fatals by cause and industry sector is only readily available for the year 1988. So Table 3 gives for 1987 some rates for industries, including vehicle accidents, but the all-industry figure is given with and without such accidents. Nevertheless, the bulk of the vehicle accidents are in the road haulage and distribution industries within the services sector. This probably accounts for services having a fatal rate not far behind that of manufacturing. The injury figures for 1988 are taken from the *Accident Yearbook 1988*.[8]

19   The comparison by rate between Spain and Great Britain is based more soundly on fatal

injuries than those all-reported. This is because of the rather wide nature of time loss injuries reported in Spain and the substantial under-reporting of over-3-day injury in Great Britain. Table 4 displays fatal injuries and rates for Spain and Great Britain.

20   Despite the inclusion of vehicle accidents in each main sector for 1987, the results for both years, 1987 and 1988, do suggest that Great Britain has a lower rate of fatality than Spain. The results for 1988 Spanish fatalities would suggest that fatal rates for the main sectors are less in Great Britain than in Spain. Another factor to be borne in mind is that some Spanish fatalities with delayed deaths may not be included while under-reporting of fatalities is reckoned not to take place in Great Britain. The comparison for agriculture is complicated by the mix of employment but the rate is above the Great Britain fatal rates based on employees, though close to rates for all in employment. For the other main industries, the Spanish rate is much higher.

21   The comparison suggests that Great Britain has a lower fatal rate over all industry than Spain and for the main industry sectors. Some of the overall difference can be explained by the differing industry mix of employees referred to in paragraph 14. Applying the Spanish industrial mix to the British fatal rates gives an overall Great Britain rate standardised for Spanish industry (at main sector level). The resulting rate is 2.39; up from 1.7

– the unstandardised published Great Britain fatal rate. So, less than 9% of the difference has been explained by industry mix.

TABLE 4 Fatal injuries and rates for Spain and Great Britain

Industry	Great Britain 1987/88		Spain 1987	
	Fatal injuries	Rate (a)	Fatal injuries	Rate (a)
Agriculture (b)	21	6.8	194	11.3
Energy (c)	33	6.7	59	44.5
Manufacturing	99	1.9	316	13.9
Construction	103	10.3	224	32.3
Services	95	0.7	389	9.0
All industries	360	1.7	1 182	14.8
All industries, excluding vehicle accidents	360	1.7		9.7
	Great Britain 1989/89		Spain 1988	
Agriculture (b)	21	7.0	149	8.8
Energy (c)	203	42.7	63	46.0
Manufacture	94	1.8	261	11.2
Construction	101	9.9	234	30.2
Services	109	0.7	265	5.8
All industry	529	2.4	972	10.2

Notes:
(a) Incidence rate per 100 000 employees
(b) For agriculture in Spain, the rate is the number of reported fatalities expressed per 100 000 in employment (employees and the self-employed). The GB rate for employees and self-employed combined is 9.4 for 1987/88 and 8.5 for 1988/89
(c) Energy comprises: coal mines, petroleum extraction and refining, electricity, gas and water. 1988/89 GB fatalities include the 167 deaths in the Piper Alpha disaster

22   The comparison, for all-industries combined, for 1987 is confirmed by the figure for 1986. The fatal rates for all industries in Spain (excluding vehicle accidents and for the three years 1986, 1987 and 1988) are: 9.9, 9.7 and 10.2. The corresponding Great Britain rates are 1.7, 1.7 and 2.4.

23   Collecting the main themes of this section together, the main point is that with allowances for vehicle accidents and the employment mix between industries, Great Britain has a lower rate of fatal injuries than Spain. A possible exception is agriculture, where British rates are similar when rates are expressed for all in employment.

Comparison by frequency rates

24   In recent years the Spanish economy has expanded in activity and hence in hours of work.[5] Spanish injury incidence rates are likely to be affected by increases in working hours. It is therefore worthwhile comparing Spain and Great

Britain in terms of fatalities related to hours worked. Frequency rates of all-reported injuries per million hours worked are published in the Spanish *Yearbook of Labour Statistics*.[4] Hours of work are also given for salaried employees and also for all in employment (employees and self-employed people).

25   An examination shows that the published frequency rates are expressed per million hours worked in a year by all those in employment. The reported injuries, however, are primarily from employees. A similar treatment was given to published injury incidence rates and induces a downward bias in rates of frequency or incidence. In this section, frequency rates of fatals are derived from hours worked by both employees and all those in employment. This will be for all industries combined and not for any individual sector.

26   The treatment of Spanish frequency rates in the Yearbook scales the average hours worked per week per person in employment by 50 working weeks per year. The resulting hours of work in 1987 is 20 950.7 millions – a figure which provides the published frequency rate.

27   The same calculation can be performed for salaried employees but will give fewer hours because employees comprise around 70% of those in employment. The hours of work and fatal frequency rates based on the two derived figures on hours worked are displayed in Table 5. Table 6 shows the comparison by frequency rate between Spain and Great Britain. The two resulting frequency rates are much higher than the Great Britain frequency rate. A further point is that the British rates are based on 41 working weeks, fewer than the 50 weeks used in deriving Spanish rates. This further strengthens the comparison where frequency rates are lower in Great Britain than in Spain.

TABLE 5 Employment, hours of work and frequency rates in Spain 1987 and 1988

Year	Worker category	Numbers of workers (thousands)	Hours worked (millions)	Frequency rate of fatal injury per million hours
1987	All in employment	11 355.4	20 950.7	0.037
	Salaried employees	7 972.9	13 912.7	0.055
1988	All in employment	11 772.7	21 779.5	0.045
	Salaried employees	8 351.5	14 698.6	0.066

TABLE 6 Frequency rate of fatal injury expressed per million hours worked: Great Britain and Spain

Great Britain (a)	Year	Spain (b)	Year
0.012	1987/88	0.037-0.055	1987
0.018	1988/89	0.045-0.066	1988

Notes:
(a) Frequency rate per million hours worked and based on a 41-week working year
(b) Frequency rate per million hours worked and based on a 50-week working year. Fatal injuries exclude deaths resulting from contact with vehicles. The lower rate is the number of reported deaths expressed per million hours worked by all in employment. The higher rate is expressed per million hours worked by salaried employees

Recent trends

28   The available figures for all industries provide separate road vehicle accidents but these cannot be separately identified for individual industries. Thus road vehicle accidents can be excluded from all-industry trends but not from trends in main industry sectors. Nevertheless this does not appear likely to affect the picture. Rates of fatal and of serious (ie fatal and severe) injury (see Table 7) rose during the latter part of the 1980s, coinciding with strong employment growth resulting from a rapid expansion of the Spanish economy (see Table 9).

29   One exception is agriculture where employment has fallen in recent years. The fatal injury rate for agriculture dropped by 8% between 1986 and 1988 (from 12.3 per 100 000 employees to 10.0) but then rose to the 1986 level in 1989. The serious injury rate increased between 1986 and 1988 but at 5% is the smallest rise of any main sector. The serious injury rate for construction rose 14% between 1986 and 1988, for other

TABLE 7 Fatal and serious injury rates 1986-1989 for all industries except the public sector compared with those for Great Britain

	Fatal		Serious		
	Spain (a)	Great Britain (a)	Spain (a)	(b)	Great Britain (a)
1986	9.9	2.0	132	137	100
1987	9.7	1.9	132	142	97
1988	10.2	2.8	151	164	96
1989	10.2p	1.9		150p	95
Change	%	%	%	%	%
1986-88	+3	+40	+14	+20	-4
1986-89	+3	-5		+9	-4

Notes:
All rates are expressed per 100 000 employees
Those for Great Britain are for planning years beginning 1 April
(a) exlcuding vehicle accidents
(b) including vehicle accidents
p = provisional

production industries (ie energy and manufacturing) it rose 20%, while the rate for services rose 31%.

TABLE 8 Fatal and serious injury rates in construction 1986 to 1989, Spain and Great Britain

	Fatal		Serious	
	Spain	Great Britain	Spain	Great Britain
1986	35.1	10.2	294	293
1987	32.0	10.3	311	287
1988	39.0	9.9	334	296
1989	36.5	9.4	311p	308
Change	%	%	%	%
1986-88	+11	-3	+14	+1

Notes:
Rates are per 100 000 employees in Great Britain, excluding vehicle accidents, for planning years 1 April – 31 March, and per 100 000 insured employees in Spain, including vehicle accidents
p = provisional

TABLE 9 Change in numbers employed by sector between 1986 and 1988, Spain and Great Britain

	Spain %	Great Britain %
Agriculture	- 1	- 5
Energy and manufacturing Energy Manufacturing	+ 5	No change - 9 + 1
Construction	+27	+ 5
All services: - transport - other services	+10	+ 6 + 4 + 7

Conclusion

30   There are a number of qualifying issues on the comparison between Spain and Great Britain. This should be based on fatal and not on non-fatal injuries. Differential reporting of fatal injuries should not greatly affect the comparison because fatal injuries are fully reported in both countries and over-reporting is likely to be negligible as well. It is not clear how over-reporting is checked in Spain but the employers and insurers alike have a positive disincentive to connive with false reporting.

31   The derivations of fatal rates, incidence and frequency, suggest that rates are lower than in Spain. This applies when expressing fatal rates both in terms of employment and in terms of hours worked. However, it is clear that the comparison shows that the rate, however expressed, is lower in Great Britain than in Spain.

32   Rates of fatal and serious injury have risen sharply during the latter part of the 1980s, both for all-industries combined and for most main industrial sectors, coinciding with strong employment growth resulting from a rapid expansion of the Spanish economy.

Bibliography

1   Interim report by Roger Clarke to European Commission working party 18 July 1990, Luxembourg

2   Private communication with Senor Manuel Davila, Subdireccion General de Estadistica, Ministerio de Trabajo y Seguridad Social, Madrid July 1990

3   Private discussion with Mr Clarke, 18 June 1990

4   *Yearbook of Labour Statistics 1988: (Anuario de Estadisticos Laborales 1988)* Spain, Ministerio de Trabajo y Seguridad Social, Madrid 1988

5   *Accidents at Work 1977-1987 (Siniestralidad Laboral 1977-1987)* Spain, Ministerio de Trabajo y Seguridad Social, Madrid 1988

6   *Bulletin of Labour Statistics No 70/1990 (Boletin de Estadisticas Laborales No 70/1990)* Ministerio de Trabajo y Seguridad Social

7   *How occupational accidents and diseases are reported in the European Community* European Foundation for the Improvement of Living and Working Conditions, Dublin 1988, ISBN 9 28 257575 6

8   *Work Accident Statistics 1988 (Estadistica de Accidentes de Trabajo 1988)* Department of Information and Statistics (Direccion General de Information y Estadistica) Ministry of Labour and Social Security (Ministerio de Trabajo y Seguridad Social) 1990

9   Private communication 11 January 1991, Senora Maria T. G. Nunez, Subdepartment of Statistics Ministry of Labour and Social Security

Appendix I –

Industrial classification

The branches of industrial activity in Spain are given in Table A1. Labour and other statistics are published on the basis of this classification.

TABLE A1 Branches of industrial activity in Spain

Branches	GB industry equivalents
Agriculture Livestock and agricultural services Forestry and game Fishing	Agriculture
Coal mines Petroleum extraction and refinery Electricity, gas and water	Energy
Extraction of minerals Basic metals Production of non-metallic minerals Chemical industry Metal processing Machinery and mechanical equipment Machinery and electrical equipment Electrical equipment and office machinery Motor vehicles Other transport equipment Precision and optical instruments Food, drink and tobacco Textile industry Leather industry Footwear, clothing and other products Wood and cork Paper and graphic arts Rubber and plastic processing Other manufacturing industries	Manufacturing
Construction	Construction
Retail trade Other trade and reprocessing Hostelry Repairs Rail transport Other land transport Sea and air transport Associated activities and communication Banking, insurance and property Leasing company services Public administration and diplomatic service Sanitary and similar Education and research Health and veterinary services Social services, cultural services Personal services Domestic services	Services

TABLE A2 Distribution of employees by main economic sector 1987

Main sector	GB %	Spain %
Agriculture	1.4	6.8
Energy	2.3	1.7
Manufacturing	23.9	28.5
Construction	4.7	8.7
Services	67.7	54.4
All industries	21 356.2 (000s)	7 972.9 (000s)